Student Resource and Solutions Manual

A First Course in Differential Equations

TENTH EDITION

Dennis G. Zill
Loyola Marymount University

Prepared by

Dennis G. Zill
Loyola Marymount University

Warren S. Wright
Loyola Marymount University

Carol D. Wright

BROOKS/COLE
CENGAGE Learning·

Australia • Brazil • Japan • Korea • Mexico • Singapore • Spain • United Kingdom • United States

BROOKS/COLE
CENGAGE Learning·

For product information and technology assistance, contact us at **Cengage Learning Customer & Sales Support, 1-800-354-9706**

For permission to use material from this text or product, submit all requests online at **www.cengage.com/permissions** Further permissions questions can be emailed to **permissionrequest@cengage.com**

ISBN-13: 978-1-133-49192-7
ISBN-10: 1-133-49192-8

Brooks/Cole
20 Channel Center Street
Boston, MA 02210
USA

Cengage Learning is a leading provider of customized learning solutions with office locations around the globe, including Singapore, the United Kingdom, Australia, Mexico, Brazil, and Japan. Locate your local office at: **www.cengage.com/global**

Cengage Learning products are represented in Canada by Nelson Education, Ltd.

To learn more about Brooks/Cole, visit **www.cengage.com/brookscole**

Purchase any of our products at your local college store or at our preferred online store **www.cengagebrain.com**

Printed in the United States of America
1 2 3 4 5   17 16 15 14 13

Table of Contents

SRSM Introduction

This *Student Resource and Solutions Manual* is meant to accompany the text *A First Course in Differential Equations with Modeling Applications*, Ninth Edition. It is intended to serve several purposes:

- to provide a list of the terminology, concepts, and skills associated with each section in the text

- to provide further elaboration on some of the material discussed in the section

- to review some of the mathematics pertinent to the material being presented (such as integration, synthetic division, and partial fractions)

- to furnish extra examples of the material presented in the section

- to illustrate the use of the computer algebra systems *Mathematica* and *Maple* in dealing with the material in the section

- to provide solutions for every third problem in the exercise sets, excluding Discussion Problems, Mathematical Models, Computer Lab Assignments, Project Problems, and Contributed Problems

- to provide hints and suggestions for selected problems throughout the entire set of exercises in each section

- to provide examples illustrating how to solve some of the Computer Lab problems.

1 INTRODUCTION TO DIFFERENTIAL EQUATIONS

The terminology and concepts listed below provide an outline of the main ideas encountered in this section. These can be useful when preparing for a quiz or test.

Terminology and Concepts

- ordinary DE

- partial DE

- order of a DE

- normal form of a DE

- linear DE

- explicit and implicit form of a solution of a DE

- interval of definition of a solution

- trivial solution of a DE

- solution curve

- family of solutions of a DE

- n-parameter family of solutions of an nth-order DE

- particular solution of a DE

- singular solution of a DE

- system of DEs

- general solution of a DE

1

The basic skills listed below summarize the more mechanical types of problems encountered in the exercise set for this section.

Basic Skills

- determine the order of a DE

- determine if a DE is linear or nonlinear

- verify that a given function is a solution of a DE

- determine the interval of definition of a solution of a DE

- use an implicit solution to determine an explict solution (this is not always possible)

- identify constant solutions of certain DEs

- verify that a pair of functions is a solution of a system of two DEs

Basic Properties of the Derivative

Definition: The derivative of a function $y = f(x)$ with respect to the independent variable x is

$$f'(x) = \lim_{h \to 0} \frac{f(x+h) - f(x)}{h},$$

whenever this limit exists.

The derivative is also denoted by y' and dy/dx. When the derivative is evaluated at $x = x_0$, denoted by $f'(x_0)$, the resulting value represents the *rate of change* of the function f at x_0. Geometrically, this is the slope of the tangent line to the graph of $y = f(x)$ at the point $(x_0, f(x_0))$. If the function f is differentiable at x_0 then the graph of f at the point $(x_0, f(x_0))$ does not have a sharp corner or vertical tangent line at that point.

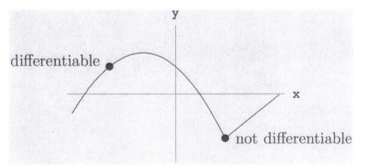

In DEs the independent variable frequently represents time and is denoted by t. In this case, if the position at time t of an object moving on a straight line is given by $s = s(t)$, then the velocity of the object at time t is the first derivative, $v(t) = s'(t)$, and the acceleration is the second derivative of $s(t)$, $a(t) = s''(t)$.

If the first derivative of a function is positive on an interval, then the function is **increasing** on that interval. Correspondingly, when the derivative is negative on an interval, the function is

decreasing on the interval. Relative extrema (maxima or minima) of a function occur where the derivative is either 0 or does not exist. The second derivative of a function can be used to determine concavity. The graph of a function f is **concave up** on an interval if $f''(x) > 0$ on the interval and **concave down** on an interval if $f''(x) < 0$ on the interval.

Hyperbolic Functions It is unfortunate that many current texts and calculus instructors downplay the hyperbolic functions. These functions are important in applied mathematics and every serious student of differential equations should know their definitions, graphs, and some of their properties. In the text, solutions of DEs will be expressed in terms of hyperbolic functions when appropriate. The six hyperbolic functions are denoted by sinh, cosh, tanh, coth, sech and csch. If x denotes a real variable, the **hyperbolic sine** and **cosine** are defined in terms of exponential functions.

Definitions:
$$\sinh x = \frac{e^x - e^{-x}}{2} \quad \text{and} \quad \cosh x = \frac{e^x + e^{-x}}{2} \tag{1}$$

The remaining four hyperbolic functions are defined in a manner analogous to trigonometry. For example, the hyperbolic tangent and cotangent are, respectively,

$$\tanh x = \frac{\sinh x}{\cosh x} \quad \text{and} \quad \coth x = \frac{\cosh x}{\sinh x}.$$

The graphs of the hyperbolic sine and cosine functions, like the graphs of natural exponential, natural logarithmic, and trigonometric functions, should be memorized.

Graphs:

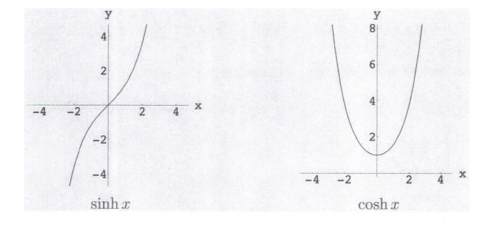

$\sinh x$ $\cosh x$

Some Properties As can be seen from both the graphs and the definitions in (1),

$$\sinh 0 = 0 \quad \text{and} \quad \cosh 0 = 1. \tag{2}$$

Also, $\cosh x$ has no real zeros and, unlike the trigonometric functions, no hyperbolic function is periodic. However, a number of other properties similar to properties for trigonometric functions do hold. As you examine the hyperbolic identities (3) through (8) below, determine whether or not the given identity differs from the analogous identity for trigonometric functions.

$$\cosh^2 x - \sinh^2 x = 1 \tag{3}$$

$$\sinh(x_1 \pm x_2) = \sinh x_1 \cosh x_2 \pm \cosh x_1 \sinh x_2 \tag{4}$$

$$\cosh(x_1 \pm x_2) = \cosh x_1 \cosh x_2 \pm \sinh x_1 \sinh x_2 \tag{5}$$

$$\sinh 2x = 2 \sinh x \cosh x \tag{6}$$

$$\cosh 2x = \cosh^2 x + \sinh^2 x \tag{7}$$

The derivatives of $\sinh x$ and $\cosh x$, given below, can be obtained using the definitions in (1).

$$\frac{d}{dx} \sinh x = \cosh x \qquad \text{and} \qquad \frac{d}{dx} \cosh x = \sinh x \tag{8}$$

Use of Computers   To define a function, such as $f(x) = e^{x^2} + \sin 2x$, use the syntax

Clear[f]   (*Mathematica*)

f[x_]:= Exp[x^2]+Sin[2x]

f[x]

```
f:=x->exp(x^2)+sin(2*x);
```    (*Maple*)

or

```
f:=unapply(exp(x^2)+sin(2*x), x);
```

If you are not familiar with *Maple* the **unapply** function probably seems like an awkward way to define a function. We will see later that it is a useful command when solving DEs.

Piecewise-defined functions are, as the name suggests, defined in pieces. For example, the function

$$f(x) = \begin{cases} x^2 + 1, & x \le 2 \\ \sin x, & x > 2 \end{cases}$$

is defined in a CAS using:

Clear[f]   (*Mathematica*)

f[x_]:= x^2 + 1/; x <= 2

f[x_]:= Sin[x]/; x > 2

```
f:=x->piecewise(x<=2, x^2+1, x>2,sin(x));
```    (*Maple*)

The syntax for the first derivative, dy/dx, of $y = f(x)$ is:

D[f[x], x]   or   f'[x]   (*Mathematica*)

```
diff(f(x), x);
```
(Maple)

The second derivative, d^2y/dx^2, of $y = f(x)$ is:

$$\mathbf{D}[f[x], \{x,2\}] \quad \text{or} \quad f''[x]$$
(Mathematica)

```
diff(f(x), x$2); or diff(f(x),x,x);
```
(Maple)

1.1 Definitions and Terminology

3. Fourth order; linear

6. Second order; nonlinear because of R^2

9. Writing the boundary-value problem in the form $x(dy/dx) + y^2 = 1$, we see that it is nonlinear in y because of y^2. However, writing it in the form $(y^2 - 1)(dx/dy) + x = 0$, we see that it is linear in x.

12. From $y = \frac{6}{5} - \frac{6}{5}e^{-20t}$ we obtain $dy/dt = 24e^{-20t}$, so that

$$\frac{dy}{dt} + 20y = 24e^{-20t} + 20\left(\frac{6}{5} - \frac{6}{5}e^{-20t}\right) = 24.$$

15. The domain of the function, found by solving $x + 2 \geq 0$, is $[-2, \infty)$. From $y' = 1 + 2(x+2)^{-1/2}$ we have

$$\begin{aligned}
(y - x)y' &= (y - x)[1 + (2(x+2)^{-1/2}] \\
&= y - x + 2(y - x)(x+2)^{-1/2} \\
&= y - x + 2[x + 4(x+2)^{1/2} - x](x+2)^{-1/2} \\
&= y - x + 8(x+2)^{1/2}(x+2)^{-1/2} = y - x + 8.
\end{aligned}$$

An interval of definition for the solution of the differential equation is $(-2, \infty)$ because y' is not defined at $x = -2$.

18. The function is $y = 1/\sqrt{1 - \sin x}$, whose domain is obtained from $1 - \sin x \neq 0$ or $\sin x \neq 1$. Thus, the domain is $\{x \mid x \neq \pi/2 + 2n\pi\}$. From $y' = -\frac{1}{2}(1 - \sin x)^{-3/2}(-\cos x)$ we have

$$2y' = (1 - \sin x)^{-3/2}\cos x = [(1 - \sin x)^{-1/2}]^3\cos x = y^3\cos x.$$

An interval of definition for the solution of the differential equation is $(\pi/2, 5\pi/2)$. Another interval is $(5\pi/2, 9\pi/2)$ and so on.

21. Differentiating $P = c_1 e^t / (1 + c_1 e^t)$ we obtain

$$\frac{dP}{dt} = \frac{(1 + c_1 e^t) c_1 e^t - c_1 e^t \cdot c_1 e^t}{(1 + c_1 e^t)^2} = \frac{c_1 e^t}{1 + c_1 e^t} \frac{[(1 + c_1 e^t) - c_1 e^t]}{1 + c_1 e^t}$$

$$= \frac{c_1 e^t}{1 + c_1 e^t} \left[1 - \frac{c_1 e^t}{1 + c_1 e^t} \right] = P(1 - P).$$

24. From $y = c_1 x^{-1} + c_2 x + c_3 x \ln x + 4x^2$ we obtain

$$\frac{dy}{dx} = -c_1 x^{-2} + c_2 + c_3 + c_3 \ln x + 8x,$$

$$\frac{d^2 y}{dx^2} = 2c_1 x^{-3} + c_3 x^{-1} + 8,$$

and

$$\frac{d^3 y}{dx^3} = -6c_1 x^{-4} - c_3 x^{-2},$$

so that

$$x^3 \frac{d^3 y}{dx^3} + 2x^2 \frac{d^2 y}{dx^2} - x \frac{dy}{dx} + y = (-6c_1 + 4c_1 + c_1 + c_1)x^{-1} + (-c_3 + 2c_3 - c_2 - c_3 + c_2)x$$

$$+ (-c_3 + c_3)x \ln x + (16 - 8 + 4)x^2$$

$$= 12x^2.$$

27. From $y = e^{mx}$ we obtain $y' = m e^{mx}$. Then $y' + 2y = 0$ implies

$$m e^{mx} + 2e^{mx} = (m + 2)e^{mx} = 0.$$

Since $e^{mx} > 0$ for all x, $m = -2$. Thus $y = e^{-2x}$ is a solution.

30. From $y = e^{mx}$ we obtain $y' = m e^{mx}$ and $y'' = m^2 e^{mx}$. Then $2y'' + 7y' - 4y = 0$ implies

$$2m^2 e^{mx} + 7m e^{mx} - 4e^{mx} = (2m - 1)(m + 4)e^{mx} = 0.$$

Since $e^{mx} > 0$ for all x, $m = \frac{1}{2}$ and $m = -4$. Thus $y = e^{x/2}$ and $y = e^{-4x}$ are solutions.

In Problems 33 and 36 we substitute $y = c$ into the differential equations and use $y' = 0$ and $y'' = 0$.

33. Solving $5c = 10$ we see that $y = 2$ is a constant solution.

36. Solving $6c = 10$ we see that $y = 5/3$ is a constant solution.

1.2 | Initial-Value Problems

The terminology and concepts listed below provide an outline of the main ideas encountered in this section. These can be useful when preparing for a quiz or test.

Terminology and Concepts

- IVPs

- initial conditions

- how the interval of definition of a solution of an IVP may differ from the domain of the function in the solution

- existence and uniqueness of solutions of IVPs

The basic skills listed below summarize the more mechanical types of problems encountered in the exercise set for this section.

Basic Skills

- given a family of solutions of a DE, find a particular solution satisfying given initial conditions

- determine a region in the plane for which a given DE will have a unique solution whose graph passes through a point in the region

- match graphs of solution curves with given initial conditions

Existence and Uniqueness

Theorem 1.2.1 is local in nature, so the interval I_0 may not be very large. The words "existence" and "uniqueness" are often not clearly understood. Use of the notation such as $y = y(x)$ and saying that a solution "exists" on the interval I_0 does *not* necessarily mean that we can actually find (meaning display) an explicit or implicit solution of the DE defined on I_0. We know in theory that a solution $y = y(x)$ exists but we may have to approximate it on I_0. The word "uniqueness" basically means we cannot have *two* (or more) solutions $y = y_1(x)$ and $y = y_2(x)$ defined on I_0 such that their solution curves both pass through the initial point (x_0, y_0). In other words, on I_0 we cannot have either of the following situations:

Curves touching on I_0 Curves crossing on I_0

Since any point on the graph of a solution curve may be considered as an initial condition of the DE resulting in that curve, no two solution curves can intersect.

3. Letting $x = 2$ and solving $1/3 = 1/(4+c)$ we get $c = -1$. The solution is $y = 1/(x^2 - 1)$. This solution is defined on the interval $(1, \infty)$.

6. Letting $x = 1/2$ and solving $-4 = 1/(1/4 + c)$ we get $c = -1/2$. The solution is $y = 1/(x^2 - 1/2) = 2/(2x^2 - 1)$. This solution is defined on the interval $(-1/\sqrt{2}, 1/\sqrt{2})$.

9. From the initial conditions we obtain

$$\frac{\sqrt{3}}{2} c_1 + \frac{1}{2} c_2 = \frac{1}{2}$$

$$-\frac{1}{2} c_1 + \frac{\sqrt{3}}{2} c_2 = 0.$$

Solving, we find $c_1 = \sqrt{3}/4$ and $c_2 = 1/4$. The solution of the initial-value problem is

$$x = (\sqrt{3}/4) \cos t + (1/4) \sin t.$$

12. From the initial conditions we obtain

$$ec_1 + e^{-1} c_2 = 0$$

$$ec_1 - e^{-1} c_2 = e.$$

Solving, we find $c_1 = \frac{1}{2}$ and $c_2 = -\frac{1}{2} e^2$. The solution of the initial-value problem is

$$y = \frac{1}{2} e^x - \frac{1}{2} e^2 e^{-x} = \frac{1}{2} e^x - \frac{1}{2} e^{2-x}.$$

15. Two solutions are $y = 0$ and $y = x^3$.

18. For $f(x,y) = \sqrt{xy}$ we have $\partial f / \partial y = \frac{1}{2} \sqrt{x/y}$. Thus, the differential equation will have a unique solution in any region where $x > 0$ and $y > 0$ or where $x < 0$ and $y < 0$.

21. For $f(x,y) = x^2/(4 - y^2)$ we have $\partial f / \partial y = 2x^2 y/(4 - y^2)^2$. Thus the differential equation will have a unique solution in any region where $y < -2$, $-2 < y < 2$, or $y > 2$.

24. For $f(x,y) = (y+x)/(y-x)$ we have $\partial f / \partial y = -2x/(y-x)^2$. Thus the differential equation will have a unique solution in any region where $y < x$ or where $y > x$.

27. Since $(2, -3)$ is not in either of the regions defined by $y < -3$ or $y > 3$, there is no guarantee of a unique solution through $(2, -3)$.

30. (a) Since $\dfrac{d}{dx} \tan(x + c) = \sec^2(x + c) = 1 + \tan^2(x + c)$, we see that $y = \tan(x + c)$ satisfies the differential equation.

(b) Solving $y(0) = \tan c = 0$ we obtain $c = 0$ and $y = \tan x$. Since $\tan x$ is discontinuous at $x = \pm \pi/2$, the solution is not defined on $(-2, 2)$ because it contains $\pm \pi/2$.

(c) The largest interval on which the solution can exist is $(-\pi/2, \pi/2)$.

33. (a) Differentiating $3x^2 - y^2 = c$ we get $6x - 2yy' = 0$ or $yy' = 3x$.

(b) Solving $3x^2 - y^2 = 3$ for y we get

$$y = \phi_1(x) = \sqrt{3(x^2 - 1)}, \qquad 1 < x < \infty,$$
$$y = \phi_2(x) = -\sqrt{3(x^2 - 1)}, \qquad 1 < x < \infty,$$
$$y = \phi_3(x) = \sqrt{3(x^2 - 1)}, \qquad -\infty < x < -1,$$
$$y = \phi_4(x) = -\sqrt{3(x^2 - 1)}, \qquad -\infty < x < -1.$$

(c) Only $y = \phi_3(x)$ satisfies $y(-2) = 3$.

36. The graph satisfies the conditions in (e).

In Problems 39 and 42 $y = c_1 \cos 2x + c_2 \sin 2x$ is a two parameter family of solutions of the second-order differential equation $y'' + 4y = 0$. In some of the problems we will use the fact that $y' = -2c_1 \sin 2x + 2c_2 \cos 2x$.

39. From the boundary conditions $y(0) = 0$ and $y\left(\dfrac{\pi}{4}\right) = 3$ we obtain

$$y(0) = c_1 = 0$$

$$y\left(\frac{\pi}{4}\right) = c_1 \cos\left(\frac{\pi}{2}\right) + c_2 \sin\left(\frac{\pi}{2}\right) = c_2 = 3.$$

Thus, $c_1 = 0$, $c_2 = 3$, and the solution of the boundary-value problem is $y = 3\sin 2x$.

42. From the boundary conditions $y(0) = 1$ and $y'(\pi) = 5$ we obtain

$$y(0) = c_1 = 1$$

$$y'(\pi) = 2c_2 = 5.$$

Thus, $c_1 = 1$, $c_2 = \dfrac{5}{2}$, and the solution of the boundary-value problem is $y = \cos 2x + \dfrac{5}{2} \sin 2x$.

1.3 Differential Equations as Mathematical Models

The terminology and concepts listed below provide an outline of the main ideas encountered in this section. These can be useful when preparing for a quiz or test.

Terminology and Concepts

- construction of mathematical models
- level of resolution of a model
- use of assumptions and empirical laws
- steps of the modeling process
- model for population growth
- model for radioactive decay
- Newton's law of cooling/warming
- model for the spread of a disease
- model for the draining of a tank
- models for first- and second-order chemical reactions
- models to describe mixtures of differing concentrations of elements in liquid solutions
- models for electrical series circuits
- models for the motion of falling bodies involving Newton's laws of motion
- approaches to DEs: analytical, qualitative, numerical

The basic skills listed below summarize the more mechanical types of problems encountered in the exercise set for this section.

Basic Skills

- set up DEs, sometimes with initial conditions, that can be used to model the situations listed above

Mixtures   In setting up DEs such as

$$\frac{dA}{dt} + \frac{1}{100} A = 6 \qquad \text{[(8) in the text]}$$

it is helpful to keep in mind the units. The derivative, dA/dt, is the rate at which the amount of salt (pounds) is changing in time (minutes) and so is measured in pounds per minute. In the derivation of (8) in the text observe that the unit "gal" cancels out in both R_{in} and R_{out} to give "lb/min." Also, note carefully that even though the tank contains 300 gallons of brine, the initial condition associated with (8) is *not* $A(0) = 300$ since $A(t)$, which represents the amount of salt at time t, is measured in pounds, whereas 300 is a measure of the number of gallons of fluid in the tank. Indeed, an initial condition for $A(t)$ is not specified. See the text: Problem 9 in Exercises 1.3.

Population Dynamics

3. Let b be the rate of births and d the rate of deaths. Then $b = k_1 P$ and $d = k_2 P^2$. Since $dP/dt = b - d$, the differential equation is $dP/dt = k_1 P - k_2 P^2$.

Newton's Law of cooling/Warming

6. By inspecting the graph in the text we take T_m to be $T_m(t) = 80 - 30 \cos \pi t/12$. Then the temperature of the body at time t is determined by the differential equation

$$\frac{dT}{dt} = k \left[T - \left(80 - 30 \cos \frac{\pi}{12} t \right) \right], \quad t > 0.$$

Mixtures

9. The rate at which salt is leaving the tank is

$$R_{out} \; (3 \text{ gal/min}) \left(\frac{A}{300} \text{ lb/gal} \right) = \frac{A}{100} \text{ lb/min}.$$

Thus $dA/dt = -A/100$ (where the minus sign is used since the amount of salt is decreasing). The initial amount is $A(0) = 50$.

12. The rate at which salt is entering the tank is

$$R_{in} = (c_{in} \text{ lb/gal})(r_{in} \text{ gal/min}) = c_{in} r_{in} \text{ lb/min}.$$

Now let $A(t)$ denote the number of pounds of salt and $N(t)$ the number of gallons of brine in the tank at time t. The concentration of salt in the tank as well as in the outflow is $c(t) = x(t)/N(t)$. But the number of gallons of brine in the tank remains steady, is increased, or is decreased depending on whether $r_{in} = r_{out}$, $r_{in} > r_{out}$, or $r_{in} < r_{out}$. In any case, the number of gallons of brine in the tank at time t is $N(t) = N_0 + (r_{in} - r_{out})t$. The output rate of salt is then

$$R_{out} = \left(\frac{A}{N_0 + (r_{in} - r_{out})t} \text{ lb/gal} \right) (r_{out} \text{ gal/min}) = r_{out} \frac{A}{N_0 + (r_{in} - r_{out})t} \text{ lb/min}.$$

The differential equation for the amount of salt, $dA/dt = R_{in} - R_{out}$, is

$$\frac{dA}{dt} = c_{in} r_{in} - r_{out} \frac{A}{N_0 + (r_{in} - r_{out})t} \quad \text{or} \quad \frac{dA}{dt} + \frac{r_{out}}{N_0 + (r_{in} - r_{out})t} A = c_{in} r_{in}.$$

Series Circuits

15. Since $i = dq/dt$ and $L\,d^2q/dt^2 + R\,dq/dt = E(t)$, we obtain $L\,di/dt + Ri = E(t)$.

Newton's Second Law and Archimedes' Principle

18. Since the barrel in Figure 1.3.17(b) in the text is submerged an additional y feet below its equilibrium position the number of cubic feet in the additional submerged portion is the volume of the circular cylinder: $\pi \times (\text{radius})^2 \times \text{height}$ or $\pi(s/2)^2 y$. Then we have from Archimedes' principle

$$\text{upward force of water on barrel} = \text{weight of water displaced}$$
$$= (62.4) \times (\text{volume of water displaced})$$
$$= (62.4)\pi(s/2)^2 y = 15.6\pi s^2 y.$$

It then follows from Newton's second law that

$$\frac{w}{g}\frac{d^2y}{dt^2} = -15.6\pi s^2 y \qquad \text{or} \qquad \frac{d^2y}{dt^2} + \frac{15.6\pi s^2 g}{w}y = 0,$$

where $g = 32$ and w is the weight of the barrel in pounds.

Newton's Second Law and Rocket Motion

21. Since the positive direction is taken to be upward, and the acceleration due to gravity g is positive, (14) in Section 1.3 becomes

$$m\frac{dv}{dt} = -mg - kv + R.$$

This equation, however, only applies if m is constant. Since in this case m includes the variable amount of fuel we must use (17) in Exercises 1.3:

$$F = \frac{d}{dt}(mv) = m\frac{dv}{dt} + v\frac{dm}{dt}.$$

Thus, replacing $m\,dv/dt$ with $m\,dv/dt + v\,dm/dt$, we have

$$m\frac{dv}{dt} + v\frac{dm}{dt} = -mg - kv + R \qquad \text{or} \qquad m\frac{dv}{dt} + v\frac{dm}{dt} + kv = -mg + R.$$

Newton's Second Law and the Law of Universal Gravitation

24. The gravitational force on m is $F = -kM_r m/r^2$. Since $M_r = 4\pi\delta r^3/3$ and $M = 4\pi\delta R^3/3$ we have $M_r = r^3 M/R^3$ and

$$F = -k\frac{M_r m}{r^2} = -k\frac{r^3 Mm/R^3}{r^2} = -k\frac{mM}{R^3}r.$$

Now from $F = ma = d^2 r/dt^2$ we have

$$m\frac{d^2 r}{dt^2} = -k\frac{mM}{R^3}r \quad \text{or} \quad \frac{d^2 r}{dt^2} = -\frac{kM}{R^3}r.$$

Additional Mathematical Models

27. The differential equation is $x'(t) = r - kx(t)$ where $k > 0$.

1.R | Chapter 1 in Review

3. $\dfrac{d}{dx}(c_1 \cos kx + c_2 \sin kx) = -kc_1 \sin kx + kc_2 \cos kx;$

$\dfrac{d^2}{dx^2}(c_1 \cos kx + c_2 \sin kx) = -k^2 c_1 \cos kx - k^2 c_2 \sin kx = -k^2(c_1 \cos kx + c_2 \sin kx);$

$\dfrac{d^2 y}{dx^2} = -k^2 y \quad \text{or} \quad \dfrac{d^2 y}{dx^2} + k^2 y = 0$

6. $y' = -c_1 e^x \sin x + c_1 e^x \cos x + c_2 e^x \cos x + c_2 e^x \sin x;$

$y'' = -c_1 e^x \cos x - c_1 e^x \sin x - c_1 e^x \sin x + c_1 e^x \cos x - c_2 e^x \sin x + c_2 e^x \cos x + c_2 e^x \cos x + c_2 e^x \sin x$

$= -2c_1 e^x \sin x + 2c_2 e^x \cos x;$

$y'' - 2y' = -2c_1 e^x \cos x - 2c_2 e^x \sin x = -2y; \qquad y'' - 2y' + 2y = 0$

9. b   **(12.)** a, b, d

15. The slope of the tangent line at (x, y) is y', so the differential equation is $y' = x^2 + y^2$.

18. (a) Differentiating $y^2 - 2y = x^2 - x + c$ we obtain $2yy' - 2y' = 2x - 1$ or $(2y - 2)y' = 2x - 1$.

   (b) Setting $x = 0$ and $y = 1$ in the solution we have $1 - 2 = 0 - 0 + c$ or $c = -1$. Thus, a solution of the initial-value problem is $y^2 - 2y = x^2 - x - 1$.

(c) Solving $y^2 - 2y - (x^2 - x - 1) = 0$ by the quadratic formula we get

$$y = \frac{2 \pm \sqrt{4 + 4(x^2 - x - 1)}}{2} = 1 \pm \sqrt{x^2 - x} = 1 \pm \sqrt{x(x-1)}.$$

Since $x(x-1) \geq 0$ for $x \leq 0$ or $x \geq 1$, we see that neither $y = 1 + \sqrt{x(x-1)}$ nor $y = 1 - \sqrt{x(x-1)}$ is differentiable at $x = 0$. Thus, both functions are solutions of the differential equation, but neither is a solution of the initial-value problem.

21. (a)

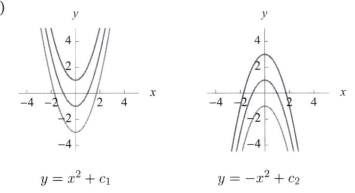

$$y = x^2 + c_1 \qquad\qquad y = -x^2 + c_2$$

(b) When $y = x^2 + c_1$, $y' = 2x$ and $(y')^2 = 4x^2$. When $y = -x^2 + c_2$, $y' = -2x$ and $(y')^2 = 4x^2$.

(c) Pasting together x^2, $x \geq 0$, and $-x^2$, $x \leq 0$, we get

$$f(x) = \begin{cases} -x^2, & x \leq 0 \\ x^2, & x > 0. \end{cases}$$

24. Differentiating $y = x \sin x + (\cos x) \ln(\cos x)$ we get

$$y' = x \cos x + \sin x + \cos x \left(\frac{-\sin x}{\cos x} \right) - (\sin x) \ln(\cos x)$$

$$= x \cos x + \sin x - \sin x - (\sin x) \ln(\cos x)$$

$$= x \cos x - (\sin x) \ln(\cos x)$$

and, $\qquad y'' = -x \sin x + \cos x - \sin x \left(\frac{-\sin x}{\cos x} \right) - (\cos x) \ln(\cos x)$

$$= -x \sin x + \cos x + \frac{\sin^2 x}{\cos x} - (\cos x) \ln(\cos x)$$

$$= -x \sin x + \cos x + \frac{1 - \cos^2 x}{\cos x} - (\cos x) \ln(\cos x)$$

$$= -x \sin x + \cos x + \sec x - \cos x - (\cos x) \ln(\cos x)$$

$$= -x \sin x + \sec x - (\cos x) \ln(\cos x).$$

Thus

$$y'' + y = -x\sin x + \sec x - (\cos x)\ln(\cos x) + x\sin x + (\cos x)\ln(\cos x) = \sec x.$$

To obtain an interval of definition we note that the domain of $\ln x$ is $(0, \infty)$, so we must have $\cos x > 0$. Thus, an interval of definition is $(-\pi/2, \pi/2)$.

27. Using implicit differentiation on $x^3 y^3 = x^3 + 1$ we have

$$3x^3 y^2 y' + 3x^2 y^3 = 3x^2$$
$$xy^2 y' + y^3 = 1$$
$$xy' + y = \frac{1}{y^2}.$$

30. Using implicit differentiation on $y = e^{xy}$ we have

$$y' = e^{xy}\left(xy' + y\right)$$
$$\left(1 - xe^x y\right)y' = ye^{xy}.$$

Since $y = e^{xy}$ we have

$$(1 - xy)y' = y \cdot y \qquad \text{or} \qquad (1 - xy)y' = y^2.$$

33. The initial conditions imply

$$c_1 e^3 + c_2 e^{-1} - 2 = 4$$
$$3c_1 e^3 - c_2 e^{-1} - 2 = -2,$$

so $c_1 = \frac{3}{2}e^{-3}$ and $c_2 = \frac{9}{2}e$. Thus $y = \frac{3}{2}e^{3x-3} + \frac{9}{2}e^{-x+1} - 2x$.

36. Figure 1.3.3 in the text can be used for reference in this problem. The differential equation is

$$\frac{dh}{dt} = -\frac{cA_0}{A_w}\sqrt{2gh}.$$

Using $A_0 = \pi(1/24)^2 = \pi/576$, $A_w = \pi(2)^2 = 4\pi$, and $g = 32$, this becomes

$$\frac{dh}{dt} = -\frac{c\pi/576}{4\pi}\sqrt{64h} = \frac{c}{288}\sqrt{h}.$$

2	FIRST-ORDER
	DIFFERENTIAL EQUATIONS

2.1	Solution Curves Without a Solution

The terminology and concepts listed below provide an outline of the main ideas encountered in this section. These can be useful when preparing for a quiz or test.

Terminology and Concepts

- lineal element
- direction field (slope field)
- autonomous first-order DE
- critical (stationary, equilibrium) point
- equilibrium solution
- phase portrait (phase line)
- asymptotically stable critical point (attractor)
- unstable critical point (repeller)
- semi-stable critical point

The basic skills listed below summarize the more mechanical types of problems encountered in the exercise set for this section.

Basic Skills

- sketch a direction field of a first-order DE by hand
- sketch a solution curve through a given point in a direction field
- find the critical points of an autonomous first-order DE
- draw the phase portrait of an autonomous first-order DE
- classify critical points as asymptotically stable (attractor), unstable (repeller) or semi-stable
- use a phase portrait to sketch solution curves of autonomous first-order DEs
- determine the behavior of solutions of autonomous first-order DEs as $t \to \pm\infty$

Signs of Derivatives   The first derivative and its algebraic sign over an interval plays an important role in the qualitative analysis in Section 2.1. If $y = f(x)$ is a differentiable function and if $dy/dx > 0$ on an interval then f is increasing on the interval, whereas if $dy/dx < 0$ on an interval then f is decreasing on the interval. Similarly, if $y = f(x)$ possesses a second derivative, then the sign of d^2y/dx^2 determines the concavity of the graph of f. If $d^2y/dx^2 > 0$ on an interval then the graph of f is concave upward on the interval, whereas if $d^2y/dx^2 < 0$ on an interval, then the graph of f is concave downward.

Use of Computers   Both *Mathematica* and *Maple* can be used to obtain direction fields of first-order DEs. In *Mathematica*, a graphics package containing the command **PlotVectorField** must first be loaded. (It is important that this be done only once during any session of *Mathematica* and that it be done before the command is first used. If this protocol is not followed, you should save your current file, quit *Mathematica*, and restart the program.) **PlotVectorField** is actually used to graph the vector field corresponding to a vector function so it must be adapted to the case of a first-order DE having the form $dy/dx = f(x, y)$. The routine below plots the direction field for the DE $dy/dx = 0.2xy$ where $-5 < x < 5$ and $-5 < y < 5$.

```
<<Graphics`                                    (Mathematica)
PlotVectorField[{1, 2x y}, {x, -5, 5}, {y, -5, 5},
    Frame->True, Axes->True, AxesLabel->{"x", "y"},
    Plotpoints->15]
```

This example will plot a direction field in which the lineal elements (arrows) have their base at a grid point. The number of grid points in both horizontal and vertical directions is specified by the **PlotPoints** option. The output is shown below in Figure 2.1. To obtain a direction field with linear elements all the same length, the graphics options **ScaleFunction** and **ScaleFactor** can be added as shown below.

```
PlotVectorField[{1, 2x y},                      (Mathematica)
    {x, -5, 5}, {y, -5, 5}, Frame->True, Axes->True,
    AxesLabel->{"x", "y"}, ScaleFunction->(0.5&),
    ScaleFactor->0.5, PlotPoints->9]
```

The output for this command is shown below in Figure 2.2.

Figure 2.1   Figure 2.2

In *Maple* the command **fieldplot** is used to obtain a direction field. The routine below plots the direction field for the DE $dy/dx = 0.2xy$.

```
with(plots);                                            (Maple)
fieldplot([1,0.2*x*y], x=-5..5, y=-5..5);
```

To obtain a direction field with lineal elements all having the same length it is necessary to scale the input by the length of each vector as shown below.

```
with(plots);                                            (Maple)
fieldplot([1/sqrt(1+(0.2*x*y)^2), 0.2*x*y/sqrt(1+(0.2*x*y)^2)],
    x=-5..5, y=-5..5);
```

2.1.1 DIRECTION FIELDS

In Problem 3 the graph corresponding to the initial condition in Part (a) *is red,* Part (b) *is green,* Part (c) *is blue, and* Part (d) *is brown. The pictures are obtain using Mathematica with*

VectorPlot[{1, f[x, y]}, {x, lhs, rhs}, {y, down, up}, . . .].

3.   **6.**

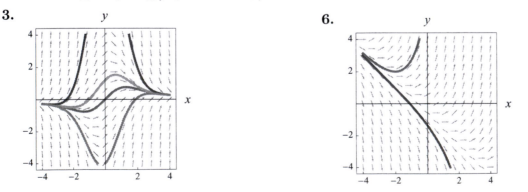

In Problems 9 and 12 the graph corresponding to the initial condition in Part (a) *is red, and* Part (b) *is blue. The pictures are obtain using Mathematica, as mentioned before Problem 1.*

9.

12.

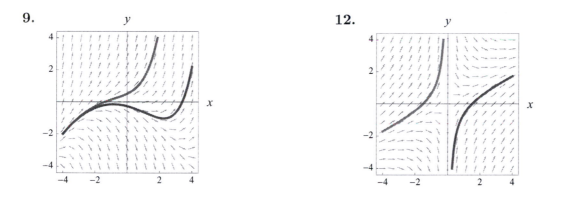

15. (a) The isoclines have the form $y = -x + c$, which are straight lines with slope -1.

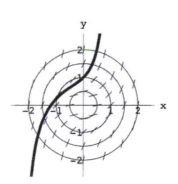

(b) The isoclines have the form $x^2 + y^2 = c$, which are circles centered at the origin.

2.1.2 AUTONOMOUS FIRST-ORDER DEs

In Problems 21–30 graphs of typical solutions are shown. However, in some of the solutions, even though the upper and lower graphs either actually bend up or down, they display as straight line segments. This is a peculiarity of the Mathematica graphing routine and may be due to the fact that the **NDSolve** *function was used rather than* **DSolve**. **NDSolve** *uses a numerical routine (see Section 2.6 in the text), and involves sampling x-coordinates where the corresponding y-coordinates are approximated. It may be that the routine involved breaks down as the graph becomes nearly vertical, forcing the x-coordinates on the graph to becomes closer and closer together.*

21. Solving $y^2 - 3y = y(y - 3) = 0$ we obtain the critical points 0 and 3. From the phase portrait we see that 0 is asymptotically stable (attractor) and 3 is unstable (repeller).

24. Solving $10 + 3y - y^2 = (5 - y)(2 + y) = 0$ we obtain the critical points -2 and 5. From the phase portrait we see that 5 is asymptotically stable (attractor) and -2 is unstable (repeller).

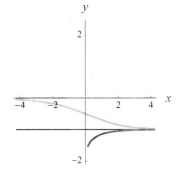

27. Solving $y \ln(y + 2) = 0$ we obtain the critical points -1 and 0. From the phase portrait we see that -1 is asymptotically stable (attractor) and 0 is unstable (repeller).

30. The critical points are approximately at -2, 2, 0.5, and 1.7. Since $f(y) > 0$ for $y < -2.2$ and $0.5 < y < 1.7$, the graph of the solution is increasing on $(-\infty, -2.2)$ and $(0.5, 1.7)$. Since $f(y) < 0$ for $-2.2 < y < 0.5$ and $y > 1.7$, the graph is decreasing on $(-2.2, 0.5)$ and $(1.7, \infty)$.

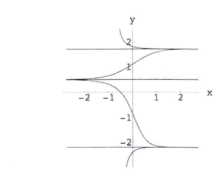

2.2 | Separable Equations

The terminology and concepts listed below provide an outline of the main ideas encountered in this section. These can be useful when preparing for a quiz or test.

Terminology and Concepts

- separable DE

- solutions with nonelementary functions

The basic skills listed below summarize the more mechanical types of problems encountered in the exercise set for this section.

Basic Skills

- determine whether or not a DE is separable

- use integration to solve a separable DE

- when possible, solve an equation for the dependent variable to find an explicit solution of a DE

Integration   The method of separation of variables is the first of several techniques for solving DEs that require integration. Recall from calculus that an indefinite integral (or antiderivative) of a function $f(x)$ is another function $F(x)$ such that $F'(x) = f(x)$. The most general indefinite integral of f is written:

$$\int f(x)dx = F(x) + c, \tag{1}$$

where $F'(x) = f(x)$ and c is an arbitrary constant. Of the many formulas given in typical tables of integrals there are only a few that appear as a matter of course in typical texts in DEs. If $u = g(x)$ is a differentiable function, then its differential is defined as $du = g'(x)dx$. Each of the following results can be verified by differentiation.

$$\int u^n du = \frac{u^{n+1}}{n+1} + c, \quad n \neq 1 \tag{2}$$

$$\int u^{-1} du = \int \frac{1}{u} du = \ln|u| + c \tag{3}$$

$$\int e^u du = e^u + c \tag{4}$$

$$\int \sin u \, du = -\cos u + c \tag{5}$$

$$\int \cos u \, du = \sin u + c \tag{6}$$

$$\int \sec^2 u \, du = \tan u + c \tag{7}$$

$$\int \csc^2 u \, du = -\cot u + c \tag{8}$$

$$\int \frac{1}{a^2 + u^2} \, du = \frac{1}{a} \tan^{-1}\left(\frac{u}{a}\right) + c \tag{9}$$

For example, to evaluate $\int \sqrt{5x - 2} \, dx$ we identify $u = 5x - 2$, $du = 5 \, dx$, and write

$$\int (5x - 2)^{1/2} \, dx = \frac{1}{5} \int \overbrace{(5x-2)^{1/2}}^{u^{1/2}} \underbrace{5 \, dx}_{du} = \frac{1}{5}\frac{2}{3}(5x-2)^{3/2} + c = \frac{2}{15}(5x-2)^{3/2} + c.$$

You are encouraged to always check your answer. In the last example, in view of the Chain Rule we have

$$\frac{d}{dx}\left[\frac{2}{15}(5x-2)^{3/2} + c\right] = \frac{2}{15}\frac{3}{2}(5x-2)^{1/2} \cdot 5 = (5x-2)^{1/2}.$$

A formula very similar to (9) uses inverse hyperbolic functions

$$\int \frac{1}{a^2 - u^2} \, du = \frac{1}{a} \tanh^{-1}\left(\frac{u}{a}\right) + c, \quad |u| < a. \tag{10}$$

Every inverse hyperbolic function is a natural logarithm. Hence, if the integrand on the left-hand side of (10) is expanded into partial fractions, then an alternative form of (10) is found to be

$$\int \frac{1}{a^2 - u^2} \, du = \frac{1}{2a} \ln\left|\frac{a + u}{a - u}\right| + c, \quad |u| \neq a. \tag{11}$$

Formulas (10) and (11) are useful in some of the problems in Exercises 3.2.

Integration by Parts   Integration by parts is a method for integrating certain products of functions as well as some inverse functions such as $\ln x$ (the inverse of e^x) and $\sin^{-1} x$. If $u = f(x)$ and $v = g(x)$, then the differentials are $du = f'(x) \, dx$ and $dv = g'(x) \, dx$, respectively. Integration by parts is a two-step procedure that begins with an integration: Integrate the function that you choose to call dv, and then differentiate the other function (by default after dv is chosen, the function remaining is u):

differentiate

$$\int u \,\boxed{dv} = \boxed{u}\,\boxed{v} - \int v \,\boxed{du}. \tag{12}$$

integrate

When choosing which portion of the integrand to designate dv you should take account of the fact that this function will need to be integrated, first to obtain v, and then integrated again to obtain

$\int v\,du$. Thus, you will normally want to choose dv both so that its integral is known to you and so that the integral of dv is no more complicated than v itself. For example, in $\int x^3 \sin x\,dx$ you want to let $dv = \sin x\,dx$ because $\sin x$ is easily integrated and $\int \sin x\,dx = -\cos x + c$ is no more complicated than $\sin x$. [If you chose $dv = x^3\,dx$, $\int x^3\,dx = \frac{1}{4}x^4 + c$ is more complicated than x^3.] There are exceptions however. For example, when finding $\int x^3 \ln x\,dx$, you should let $u = \ln x$ and $dv = x^3\,dx$.

Some Important Integrals   The following integrals require integration by parts.

(a) If $n = 1, 2, 3, \ldots$, the indefinite integral of the form $\int x^n e^{ax}\,dx$, a a constant, demands integration by parts n times. For example, to evaluate $\int x^2 e^{5x}\,dx$ we must use (12) twice. It is also important to remember that for integrals of this type, integration by parts always starts by integrating the exponential function. For example,

$$\int x\boxed{e^{-2x}}\,dx = \boxed{x}\boxed{\left(-\tfrac{1}{2}e^{-2x}\right)} - \int \boxed{1}\cdot\left(-\frac{1}{2}e^{-2x}\right)dx = -\frac{1}{2}xe^{-2x} - \frac{1}{4}e^{-2x} + c.$$

differentiate

integrate

(b) An integral $\int x^\alpha \ln x\,dx$, $\alpha \neq -1$, begins by integrating x^α. For example,

$$\int x^3 \ln x\,dx = \frac{1}{4}x^4 \ln x - \frac{1}{4}\int x^4 \frac{1}{x}\,dx = \frac{1}{4}x^4 \ln x - \frac{1}{4}\int x^3\,dx = \frac{1}{4}x^4 \ln x - \frac{1}{16}x^4 + c.$$

The special case when $\alpha = -1$, does not require integration by parts. By writing

$$\int x^{-1} \ln x\,dx = \int \ln x\,\frac{dx}{x}$$

and identifying $u = \ln x$, $du = dx/x$, the integral is recognized as the form given in (2) with $n = 1$:

$$\int x^{-1} \ln x\,dx = \int \overset{u^1}{\overbrace{\ln x}}\,\overset{du}{\overbrace{\frac{dx}{x}}} = \frac{1}{2}(\ln x)^2 + c.$$

Verify this last result by differentiating the right-hand side.

(c) You might recall that integrals of the type $\int e^{\alpha x} \sin \beta x\,dx$ or $\int e^{\alpha x} \cos \beta x\,dx$ require integration by parts twice. The unusual feature of these integrals is that in the second application of integration by parts the original integral is recovered on the right-hand side. We then solve for the original integral. It is hard to make a mistake here since we can begin the process by integrating either the exponential function or the trigonometric function. But it is important to remember that if the first integraton by parts is begun by integrating, say, the exponential function, then the second integration by parts must also begin by integrating the exponential function. For example,

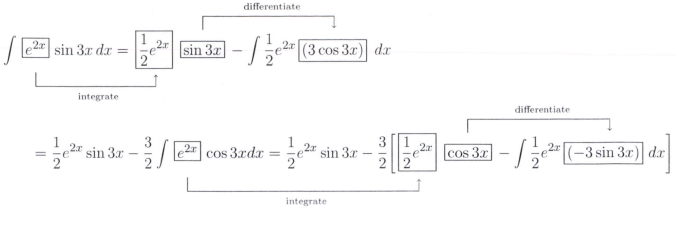

$$= \frac{1}{2}e^{2x}\sin 3x - \frac{3}{4}e^{2x}\cos 3x - \frac{9}{4}\int e^{2x}\sin 3x\, dx.$$

If the original integral is denoted by the symbol I then the last result is equivalent to

$$I = \frac{1}{2}e^{2x}\sin 3x - \frac{3}{4}e^{2x}\cos 3x - \frac{9}{4}I.$$

Solving for I then gives $\frac{13}{4}I = \frac{1}{2}e^{2x}\sin 3x - \frac{3}{4}e^{2x}\cos 3x$ or

$$I = \int e^{2x}\sin 3x\, dx = \frac{2}{13}e^{2x}\sin 3x - \frac{3}{13}e^{2x}\cos 3x + c.$$

See **Use of Computers** later in this section for *Mathematica* and *Maple* implementations of both indefinite and definite integration.

Partial Fractions The following discussion illustrates a procedure for integrating certain rational functions $P(x)/Q(x)$, where we shall assume for this discussion that the degree of $P(x)$ is less than the degree of $Q(x)$. (If the degree $P(x) \geq$ degree $Q(x)$, then start the process with long division.) This method, known as **partial fractions**, consists of expanding such a rational function into simpler component fractions, and then evaluating the integral term-by-term. We begin with an example. When the terms in the sum

$$\frac{2}{x+5} + \frac{1}{x+1} \tag{13}$$

are combined by means of a common denominator, we obtain the single rational expression

$$\frac{3x+7}{(x+5)(x+1)}. \tag{14}$$

Now suppose that we are faced with the problem of evaluating the integral $\displaystyle\int \frac{3x+7}{(x+5)(x+1)}\, dx$. Of course, the solution is obvious. We use the equality of (13) and (14) to write:

$$\int \frac{3x+7}{(x+5)(x+1)}\, dx = \int \frac{2}{x+5}\, dx + \int \frac{1}{x+1}\, dx = 2\ln|x+5| + \ln|x+1| + c.$$

In the discussion that follows, we review the assumptions and the algebra of four cases of partial fraction decomposition.

CASE 1 Nonrepeated Linear Factors

We state the following fact from algebra without proof. If

$$\frac{P(x)}{Q(x)} = \frac{P(x)}{(a_1x + b_1)(a_2x + b_2) \cdots (a_nx + b_n)}$$

where all the factors $a_ix + b_i$, $i = 2, \ldots, n$, in the denominator are distinct and the degree of $P(x)$ is less than n, then unique real constants $C_1, C_2, \ldots, C_n$ exist such that

$$\frac{P(x)}{Q(x)} = \frac{C_1}{a_1x + b_1} + \frac{C_2}{a_2x + b_2} + \cdots + \frac{C_n}{a_nx + b_n}.$$

For example, to expand $(2x + 1)/(x - 1)(x + 3)$ into individual partial fractions we make the assumption that the integrand can be written as

$$\frac{2x + 1}{(x - 1)(x + 3)} = \frac{A}{x - 1} + \frac{B}{x + 3}.$$

Combining the terms of the right-hand member of the equation over a common denominator gives

$$\frac{2x + 1}{(x - 1)(x + 3)} = \frac{A(x + 3) + B(x - 1)}{(x - 1)(x + 3)}.$$

Since the denominators are identical, the numerators are also identical

$$2x + 1 = A(x + 3) + B(x - 1) = (A + B)x + (3A - B), \tag{15}$$

and the coefficients of the powers of x are the same:

$$2 = A + B \quad \text{and} \quad 1 = 3A - B.$$

These simultaneous equations can now be solved for A and B. The results are $A = \frac{3}{4}$ and $B = \frac{5}{4}$. Therefore,

$$\frac{2x + 1}{(x - 1)(x + 3)} = \frac{\frac{3}{4}}{x - 1} + \frac{\frac{5}{4}}{x + 3}.$$

Note: The numbers A and B in the preceding example can be determined in an alternative manner. Since (15) is an identity, that is, the equality is true for every value of x, it holds for $x = 1$ and $x = -3$, *the zeros of the denominator.* Setting $x = 1$ in (15) gives $3 = 4A$, from which it follows immediately that $A = \frac{3}{4}$. Similarly, by setting $x = -3$ in (15), we obtain $-5 = (-4)B$ or $B = \frac{5}{4}$. In the *Remarks* at the end of Section 7.2 in the text an equivalent and simpler method for evaluating the coefficients in a partial fraction decomposition is given. It is important to be aware that this technique, called the **cover-up method**, works only when the denomintor is a product of linear factors, none of which are repeated.

CASE II Repeated Linear Factors

If

$$\frac{P(x)}{Q(x)} = \frac{P(x)}{(ax+b)^n}$$

where $n > 1$ and the degree of $P(x)$ is less than n, then unique real constants $C_1, C_2, \ldots, C_n$ can be found such that

$$\frac{P(x)}{(ax+b)^n} = \frac{C_1}{ax+b} + \frac{C_2}{(ax+b)^2} + \cdots + \frac{C_n}{(ax+b)^n}.$$

Combining the Cases: When the denominator $Q(x)$ contains distinct as well as repeated inear factors, we combine the two cases. For example, to expand $(6x-1)/x^3(2x-1)$ into partial fractions we write

$$\frac{6x-1}{x^3(2x-1)} = \frac{A}{x} + \frac{B}{x^2} + \frac{C}{x^3} + \frac{D}{2x-1}$$

since x^3 means that the linear term x is used as a factor x three times. Putting the right-hand side over a common denominator and equating numerators gives

$$6x - 1 = Ax^2(2x-1) + Bx(2x-1) + C(2x-1) + Dx^3 \tag{16}$$

$$= (2A + D)x^3 + (-A + 2B)x^2 + (-B + 2C)x - C. \tag{17}$$

If we set $x = 0$ and $x = \frac{1}{2}$ in (16), we find $C = 1$ and $D = 16$, respectively. Now since the denominator of the original expression has only two distinct zeros, we can find A and B by equating the coefficients of x^3 and x^2 in

$$6x - 1 = 0x^3 + 0x^2 + 6x - 1 = (2A + 16)x^3 + (-A + 2B)x^2 + (-B + 2)x - 1.$$

Thus, $A = -8$ and $B = -4$. Therefore,

$$\frac{6x-1}{x^3(2x-1)} = -\frac{8}{x} - \frac{4}{x^2} + \frac{1}{x^3} + \frac{16}{2x-1}.$$

CASE III Nonrepeated Quadratic Factors

Suppose the denominator of the rational function $P(x)/Q(x)$ can be expressed as a product of distinct quadratic factors $a_ix^2 + b_ix + c_i$, $i = 1, 2, \ldots, n$, that are irreducible (not factorable) over the real numbers. If the degree of $P(x)$ is less than $2n$, we can find unique real constants $A_1, A_2, \ldots, A_n, B_1, B_2, \ldots, B_n$ such that

$$\frac{P(x)}{(a_1x^2 + b_1x + c_1)(a_2x^2 + b_2x + c_2) \cdots (a_nx^2 + b_nx + c_n)}$$

$$= \frac{A_1x + B_1}{a_1x^2 + b_1x + c} + \frac{A_2x + B_2}{a_2x^2 + b_2x + c_2} + \cdots + \frac{A_nx + B_n}{a_nx^2 + b_nx + c_n}.$$

For example, to expand $4x/(x^2+1)(x^2+2x+3)$ into partial fractions we first write

$$\frac{4x}{(x^2+1)(x^2+2x+3)} = \frac{Ax+B}{x^2+1} + \frac{Cx+D}{x^2+2x+3}$$

and, after putting the right-hand side over a common denominator, equate numerators:

$$4x = (Ax+B)(x^2+2x+3) + (Cx+D)(x^2+1)$$

$$= (A+B)x^3 + (2A+B+D)x^2 + (3A+2B+C)x + (3B+D).$$

Since the denominator of the original fraction has no real zeros, we compare coefficients of powers of x:

$$0 = A+C$$

$$0 = 2A+B+D$$

$$4 = 3A+2B+C$$

$$0 = 3B+D.$$

Solving the equations yields $A=1$, $B=1$, $C=-1$, and $D=-3$. Therefore,

$$\frac{4x}{(x^2+1)(x^2+2x+3)} = \frac{x+1}{x^2+1} - \frac{x+3}{x^2+2x+3}.$$

CASE IV Repeated Quadratic Factors

In the case when the denominator contains a repeated irreducible quadratic polynomial ax^2+bx+c,

$$\frac{P(x)}{Q(x)} = \frac{P(x)}{(ax^2+bx+c)^n},$$

then if $P(x)$ is of degree less than $2n$, we can find real constants $A_1, A_2, \ldots, A_n, B_1, B_2, \ldots, B_n$ such that

$$\frac{P(x)}{(ax^2+bx+c)^n} = \frac{A_1x+B_1}{ax^2+bx+c} + \frac{A_2x+B_2}{(ax^2+bx+c)^2} + \cdots + \frac{A_nx+B_n}{(ax^2+bx+c)^n}.$$

Combining the Cases: The denominator $Q(x)$ in the fraction $(x+3)/x^2(x-5)(x^2+1)^2$ contains a distinct linear factor $x-5$, repeated linear factors x^2, and repeated irreducible quadratic factors $(x^2+1)^2$. To expand into partial fractions, we write

$$\frac{x+3}{x^2(x-5)(x^2+1)^2} = \underbrace{\frac{A}{x} + \frac{B}{x^2}}_{repeated\ linear} + \underbrace{\frac{C}{x-5}}_{linear} + \underbrace{\frac{Dx+E}{x^2+1} + \frac{Fx+G}{(x^2+1)^2}}_{repeated\ quadratic}.$$

Important Note: Although we have used it as a technique of integration, the ability to expand a rational expression $P(x)/Q(x)$ into partial fractions is also used in another context in the study of DEs; *it is frequently essential* when computing the inverse Laplace transform in Chapter 7. See

Section 7.2 in this manual for the syntax used in *Mathematica* and *Maple* to obtain a partial fraction decomposition of a rational function.

Use of Computers   We show below how *Mathematica* and *Maple* can be used to graph a function, graph an equation, find (and name) the solution of an equation, and find the indefinite integral (or antiderivative) and definite integral of a function.

Graphing a Function: To graph a function, such as $f(x) = \sqrt{4 - x^2}$, use the syntax

$$\textbf{Clear}[f] \hspace{8cm} (\textit{Mathematica})$$

$$\textbf{f}[\textbf{x\_}] := \textbf{Sqrt}[4 - \textbf{x}\hat{\ }2]$$

$$\textbf{f}[\textbf{x}]$$

$$\textbf{Plot}[\textbf{f}[\textbf{x}], \{\textbf{x}, -3, 3\}]$$

```
f:=x->sqrt(4-x^2);                          (Maple)
plot(f(x), x = -3..3);
```

Because $\sqrt{4 - x^2}$ is not a real number for $x < -2$ or $x > 2$, *Mathematica* returns some error messages when it tries to evaluate the function at points in these intervals; it does however then plot the graph. *Maple*, on the other hand, appears to ignore such problems and simply plots the graph. Since the graph of $f(x)$ is the upper half of the circle $x^2 + y^2 = 4$, you would expect the graph to look like a semicircle. Due, however, to the fact that both programs choose the widths and heights of their plots without regard to the scales on the horizontal and vertical axes, the semicircle may look more like a semi-ellipse. To insure that one unit on each axis has the same length on the computer screen, a graphics option is used as shown below.

$$\textbf{Plot}[\textbf{f}[\textbf{x}], \{\textbf{x}, -2, 2\}, \textbf{AspectRatio} -> 2/4] \hspace{3cm} (\textit{Mathematica})$$

```
plot(f(x), x=-3..3, scaling=CONSTRAINED);       (Maple)
```

(The **AspectRatio** option in the **Plot** command is the height of the plot divided by its width.)

Both programs provide a large number of graphics options that can be used to control the appearance of plots. Probably the most commonly used of these is an option that specifies the extent of the vertical axis to be shown. This may be necessary to see details of the graph that are obscured if the CAS (as opposed to the user) is allowed to choose the range of the graph shown. Examples are shown below with pictures displayed for the *Mathematica* output. The pictures generated by *Maple* are similar.

$$\textbf{Plot}[(\textbf{x}\hat{\ }2 - 1)\textbf{Exp}[2\textbf{x}\hat{\ }2], \{\textbf{x}, -2, 2\}, \hspace{3cm} (\textit{Mathematica})$$

$$\textbf{PlotRange} -> \{-2, 2\}]$$

```
plot((x^2-1)*exp(2*x^2), x = -2..2,          (Maple)
    y = -2..2));
```

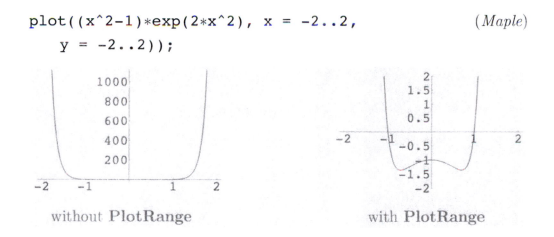

without **PlotRange**   with **PlotRange**

Finally, you may want to superimpose the graphs of two or more functions on the same set of coordinate axes. The commands below show the graphs of $f(x) = x^3 - 3x^2 + x + 1$ and $g(x) = 1 - 5x/4$, the tangent line to the graph of $f(x)$ at $x = 3/2$.

```
Clear[f, g]                                  (Mathematica)
f[x_]:= x^3 - 3x^2 + x + 1
g[x_]:= 1 - 5x/4
{f[x], g[x]}
Plot[{f(x), g(x)}, {x, -2, 3}]
```

```
f:=x->x^3-3*x^2+x+1;                          (Maple)
g:=x->1-5*x/4;
plot({f(x),g(x)}, x = -2..3);
```

Graphing an Equation: In addition to graphing functions, a CAS can be used to graph equations that implicitly define functions. Equations such as this frequently arise when solving a DE using separation of variables. For example, the solution of $(1 + e^y)y' = x \cos x + \sin x$, $y(0) = 0$, obtained using separation of variables, is $y + e^y = x \sin x + 1$. We cannot solve this equation for y in terms of elementary functions, but we can graph the equation using **ContourPlot** in *Mathematica* and **contourplot** in *Maple* as shown below.

```
f[x_, y_]:= y + Exp[y] - x Sin[x] - 1         (Mathematica)
f[x, y]
ContourPlot[f[x, y], {x, -15, 15}, {y, -10, 5}, PlotPoints->50,
    Contours->{0}, ContourShading->False, Frame->False,
    Axes->True, AxesOrigin->{0, 0}, AxesLabel->{"x", "y"}]
```

Many of the graphics options used above, such as **Frame** and **Axes** are self-explanatory. The default setting for **PlotPoints** is 15 and may result in a graph consisting of straight line segments and sharp corners. The **Contours** options is set to 0 in this case because the equation being graphed is $f(x, y) = y + e^y - x \sin x - 1 = 0$. Alternatively, the function could have been specified as $f(x, y) = y + e^x - x \sin x$ and the **Contours** option set to 1.

In *Maple* we effect the same result using the commands

```
f:=(x,y)->y+exp(y)-x*sin(x)-1;                    (Maple)
with(plots);
contourplot(f(x,y), x=-15..15, y=-10..5,
    contours=[0], grid=[80,80]);
```

The `grid` option is set to numbers that are large enough to insure a fairly smooth-looking graph.

Solving an Equation: Depending on the nature of an equation, there are a number of commands in both *Mathematica* and *Maple* that can be used to find solutions of the equation. If the equation involves only polynomial or rational functions, use **Solve** or **NSolve** in *Mathematica* and **solve** in *Maple*.

$$\text{Solve}[2\text{x}^2 - 3\text{x} - 4 == 0, \text{x}] \qquad\qquad (Mathematica)$$

or

$$\text{NSolve}[2\text{x}^2 - 3\text{x} - 4 == 0, \text{x}]$$

and

```
solve(2*x^2-3*x-4=0, x);                          (Maple)
```

Both **Solve** and **solve** will give exact solutions when possible. To obtain decimal versions of the solutions use **NSolve** in *Mathematica* and insert a decimal point in one of the numbers in the equation (`2*x^2-3*x-4.=0`) in *Maple*. (This will also work with the **Solve** command in *Mathematica* as an alternative to **NSolve**.)

To solve equations involving more complicated expressions such as $e^x = 3 \cos x + 1$ use **FindRoot** in *Mathematica* and **fsolve** in *Maple*. In both programs, either Newton's method or a variation of it is used, so some form of an initial guess should be provided. This can be obtained from graphs and will enable you to select the solution you want in the event there is more than one.

$$\text{Plot}[\{\text{Exp}[\text{x}], 3\text{Cos}[\text{x}]+1\}, \{\text{x, -5, 5}\}] \qquad\qquad (Mathematica)$$
$$\text{FindRoot}[\text{Exp}[\text{x}]==3\text{Cos}[\text{x}]+1, \{\text{x, 1}\}]$$

```
plot([exp(x), 3*cos(x)+1], x=-5..5, y=-3..10);  (Maple)
fsolve(exp(x), 3*cos(x)+1, x=0..2);
```

As shown above, in *Maple*, a range of values can be specified. Other options are available, but this is generally the most effective way to find the solution you are looking for. If there is no solution within this range, no solution will be given. If there are multiple solutions within the range, only one will be given, unless you are solving a polynomial equation, in which case all solutions within the range will be given.

Naming the Solution of an Equation: If you intend to use the solution of an equation in a subsequent computation in your CAS, then it is convenient to have a name for this value so that you do not have to read it from the screen and then type it in. The routines below show how to do this, assigning the name a to the solution of $e^{3x-x^2} \cos x = 0$.

Clear[f]   (*Mathematica*)
f[x_]:= Exp[x - x^2] + Cos[x]
f[x]
Plot[f[x], {x, -5, 5}]
sol = FindRoot[f[x]== 0, {x, 2}]
a = x/.sol

```
f:=x->exp(x-x^2)+cos(x);                          (Maple)
plot(f(x), x=-5..5);
a:=fsolve(f(x)=0, x=1..3);
```

Integrals: *Mathematica* and *Maple* can find both antiderivatives (indefinite integrals) of many functions and definite integrals of most functions. To find an antiderivative of $f(x)$ use

Integrate[f[x], x]   (*Mathematica*)

```
int(f(x), x);                                        (Maple)
```

If you want to further work with the antiderivative in the CAS it can be helpful to define it as a function that can itself be evaluated at a point, graphed, or differentiated. The syntax to accomplish this is

g[x_]:= Evaluate[Integrate[f(x), x]]   (*Mathematica*)
g[x]

```
g:=unapply(int(f(x),x),x);                           (Maple)
```

To find the definite integral, $\int_a^b f(x)\,dx$, of f we use

$$\textbf{Integrate}[\textbf{f(x)}, \{\textbf{x, a, b}\}] \qquad\qquad (Mathematica)$$

$$\texttt{int(f(x),x=a..b);} \qquad\qquad (Maple)$$

The comands **Integrate** and `int` will try to find exact values of a definite integral by evaluating the antiderivative at the endpoints of the interval. If the CAS is unable to find an antiderivative of $f(x)$ (for example, when $f(x) = \sin(\sin x)$, or if you simply want a numerical approximation of the definite integral, use

$$\textbf{NIntegrate}[\textbf{f(x)}, \{\textbf{x, a, b}\}] \qquad\qquad (Mathematica)$$

$$\texttt{evalf(int(f(x),x=a..b));} \qquad\qquad (Maple)$$

In this section and ones following we will encounter an expression of the form $\ln|g(y)| = f(x) + c$. *To solve for* $g(y)$ *we exponentiate both sides of the equation. This yields* $|g(y)| = e^{f(x)+c} = e^c e^{f(x)}$ *which implies* $g(y) = \pm e^c e^{f(x)}$. *Letting* $c_1 = \pm e^c$ *we obtain* $g(y) = c_1 e^{f(x)}$.

3. From $dy = -e^{-3x}\,dx$ we obtain $y = \frac{1}{3}e^{-3x} + c$.

6. From $\dfrac{1}{y^2}\,dy = -2x\,dx$ we obtain $-\dfrac{1}{y} = -x^2 + c$ or $y = \dfrac{1}{x^2 + c_1}$.

9. From $\left(y + 2 + \dfrac{1}{y}\right)dy = x^2 \ln x\,dx$ we obtain $\dfrac{y^2}{2} + 2y + \ln|y| = \dfrac{x^3}{3}\ln|x| - \dfrac{1}{9}x^3 + c$.

12. From $2y\,dy = -\dfrac{\sin 3x}{\cos^3 3x}\,dx$ or $2y\,dy = -\tan 3x \sec^2 3x\,dx$ we obtain $y^2 = -\frac{1}{6}\sec^2 3x + c$.

15. From $\dfrac{1}{S}\,dS = k\,dr$ we obtain $S = ce^{kr}$.

18. From $\dfrac{1}{N}\,dN = \left(te^{t+2} - 1\right)dt$ we obtain $\ln|N| = te^{t+2} - e^{t+2} - t + c$ or $N = c_1 e^{te^{t+2} - e^{t+2} - t}$.

21. From $x\,dx = \dfrac{1}{\sqrt{1 - y^2}}\,dy$ we obtain $\frac{1}{2}x^2 = \sin^{-1} y + c$ or $y = \sin\left(\dfrac{x^2}{2} + c_1\right)$.

24. From $\dfrac{1}{y^2 - 1}\,dy = \dfrac{1}{x^2 - 1}\,dx$ or $\dfrac{1}{2}\left(\dfrac{1}{y-1} - \dfrac{1}{y+1}\right)dy = \dfrac{1}{2}\left(\dfrac{1}{x-1} - \dfrac{1}{x+1}\right)dx$ we obtain

$\ln|y - 1| - \ln|y + 1| = \ln|x - 1| - \ln|x + 1| + \ln c$ or $\dfrac{y-1}{y+1} = \dfrac{c(x-1)}{x+1}$. Using $y(2) = 2$ we find

$c = 1$. A solution of the initial-value problem is $\dfrac{y-1}{y+1} = \dfrac{x-1}{x+1}$ or $y = x$.

27. Separating variables and integrating we obtain

$$\frac{dx}{\sqrt{1-x^2}} - \frac{dy}{\sqrt{1-y^2}} = 0 \quad \text{and} \quad \sin^{-1}x - \sin^{-1}y = c.$$

Setting $x = 0$ and $y = \sqrt{3}/2$ we obtain $c = -\pi/3$. Thus, an implicit solution of the initial-value problem is $\sin^{-1}x - \sin^{-1}y = -\pi/3$. Solving for y and using an addition formula from trigonometry, we get

$$y = \sin\left(\sin^{-1}x + \frac{\pi}{3}\right) = x\cos\frac{\pi}{3} + \sqrt{1-x^2}\sin\frac{\pi}{3} = \frac{x}{2} + \frac{\sqrt{3}\sqrt{1-x^2}}{2}.$$

30. Separating variables, integrating from -2 to x, and using t as a dummy variable of integration gives

$$\int_{-2}^{x}\frac{1}{y^2}\frac{dy}{dt}\,dt = \int_{-2}^{x}\sin t^2\,dt$$

$$-y(t)^{-1}\Big|_{-2}^{x} = \int_{-2}^{x}\sin t^2\,dt$$

$$-y(x)^{-1} + y(-2)^{-1} = \int_{-2}^{x}\sin t^2\,dt$$

$$-y(x)^{-1} = -y(-2)^{-1} + \int_{-2}^{x}\sin t^2\,dt$$

$$y(x)^{-1} = 3 - \int_{-2}^{x}\sin t^2\,dt.$$

Thus

$$y(x) = \frac{1}{3 - \int_{-2}^{x}\sin t^2\,dt}.$$

33. Writing the differential equation as $e^x\,dx = e^{-y}\,dy$ and integrating we have $e^x = -e^{-y} + c$. Using $y(0) = 0$ we find that $c = 2$ so that $y = -\ln(2 - e^x)$.

To find the interval of definition of this solution we note that $2 - e^x > 0$ so x must be in $(-\infty, \ln 2)$.

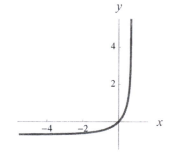

36. Separating variables, we have

$$\frac{dy}{y^2 - y} = \frac{dx}{x} \quad \text{or} \quad \int \frac{dy}{y(y-1)} = \ln|x| + c.$$

Using partial fractions, we obtain

$$\int \left(\frac{1}{y-1} - \frac{1}{y} \right) dy = \ln|x| + c$$

$$\ln|y-1| - \ln|y| = \ln|x| + c$$

$$\ln \left| \frac{y-1}{xy} \right| = c$$

$$\frac{y-1}{xy} = e^c = c_1.$$

Solving for y we get $y = 1/(1 - c_1 x)$. We note by inspection that $y = 0$ is a singular solution of the differential equation.

(a) Setting $x = 0$ and $y = 1$ we have $1 = 1/(1 - 0)$, which is true for all values of c_1. Thus, solutions passing through $(0, 1)$ are $y = 1/(1 - c_1 x)$.

(b) Setting $x = 0$ and $y = 0$ in $y = 1/(1 - c_1 x)$ we get $0 = 1$. Thus, the only solution passing through $(0, 0)$ is $y = 0$.

(c) Setting $x = \frac{1}{2}$ and $y = \frac{1}{2}$ we have $\frac{1}{2} = 1/(1 - \frac{1}{2}c_1)$, so $c_1 = -2$ and $y = 1/(1 + 2x)$.

(d) Setting $x = 2$ and $y = \frac{1}{4}$ we have $\frac{1}{4} = 1/(1 - 2c_1)$, so $c_1 = -\frac{3}{2}$ and $y = 1/(1 + \frac{3}{2}x) = 2/(2 + 3x)$.

39. The singular solution $y = 1$ satisfies the initial-value problem.

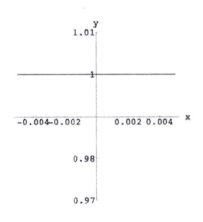

42. Separating variables we obtain $\dfrac{dy}{(y-1)^2 + 0.01} = dx$. Then

$$10\tan^{-1} 10(y-1) = x+c \quad \text{and} \quad y = 1 + \frac{1}{10}\tan\frac{x+c}{10}.$$

Setting $x = 0$ and $y = 1$ we obtain $c = 0$. The solution is

$$y = 1 + \frac{1}{10}\tan\frac{x}{10}.$$

Alternatively, we can use the fact that

$$\int \frac{dy}{(y-1)^2 - 0.01} = -\frac{1}{0.1}\tanh^{-1}\frac{y-1}{0.1} = -10\tanh^{-1} 10(y-1).$$

We use the inverse hyperbolic tangent because $|y-1| < 0.1$ or $0.9 < y < 1.1$. This follows from the initial condition $y(0) = 1$. Solving the above equation for y we get $y = 1 + 0.1\tanh(x/10)$.

45. We separate variable and rationalize the denominator:

$$dy = \frac{1}{1 + \sin x} \cdot \frac{1 - \sin x}{1 - \sin x}\, dx = \frac{1 - \sin x}{1 - \sin^2 x}\, dx = \frac{1 - \sin x}{\cos^2 x}\, dx$$

$$= (\sec^2 x - \tan x \sec x)\, dx.$$

Integrating, we have $y = \tan x - \sec x + C$.

48. Separating variables and integrating we have

$$\int \frac{dy}{y^{2/3}(1 - y^{1/3})} = \int dx$$

$$\int \frac{y^{2/3}}{1 - y^{1/3}}\, dy = x + c_1$$

$$-3\ln\left|1 - y^{1/3}\right| = x + c_1$$

$$\ln\left|1 - y^{1/3}\right| = -\frac{x}{3} + c_2$$

$$\left|1 - y^{1/3}\right| = c_3 e^{-x/3}$$

$$1 - y^{1/3} = c_4 e^{-x/3}$$

$$y^{1/3} = 1 + c_5 e^{-x/3}$$

$$y = \left(1 + c_5 e^{-x/3}\right)^3.$$

2.3 | Linear Equations

The terminology and concepts listed below provide an outline of the main ideas encountered in this section. These can be useful when preparing for a quiz or test.

Terminology and Concepts

- linear first-order DE

- homogeneous linear DE

- variation of parameters as applied to a linear first-order DE
- integrating factor

- general solution of a linear first-order DE

- singular points of a linear first-order DE

- transient term

- special functions

The basic skills listed below summarize the more mechanical types of problems encountered in the exercise set for this section.

Basic Skills

- find the integrating factor for a linear first-order DE

- use the integrating factor to find the general solution of a linear first-order DE

- find transient terms in the general solution of a linear first-order DE

- solve a linear first-order DE satisfying an initial condition

- solve a linear first-order IVP with a piecewise-defined coefficient

- determine an interval of definition of the general solution of a linear first-order DE

An Identity   In precalculus mathematics you learned that the exponential statement $e^N = M$ is equivalent to the logarithmic expression $N = \ln M$. If we substitute this last expression into the first, we obtain the important identity

$$e^{\ln M} = M,$$

which is very useful in this section when computing the integrating factor $e^{\int P(x)\, dx}$. For example, to determine the integrating factor for $(5 + x^2)\dfrac{dy}{dx} - 2xy = x$ we first put the DE into standard form $\dfrac{dy}{dx} - \dfrac{2x}{5 + x^2}\, y = \dfrac{x}{5 + x^2}$ and identify $P(x) = -2x/(5 + x^2)$:

$$e^{-\int \frac{2x}{5+x^2}\, dx} = e^{-\ln|5+x^2|} = e^{\ln|5+x^2|^{-1}} = (5 + x^2)^{-1} = \frac{1}{5 + x^2}.$$

We were able to discard the absolute value sign in $|5 + x^2|$ because $5 + x^2 > 0$ for all x.

It may take some time to get used to the fact that once the DE has been put into standard form and the integrating factor determined, multiplying both sides of the standard form by the integrating factor recasts the left-hand side of the equation into the derivative of the integrating factor and the dependent variable. Above, we saw that the integrating factor for $\dfrac{dy}{dx} - \dfrac{2x}{5 + x^2} y = \dfrac{x}{5 + x^2}$ is $\dfrac{1}{5 + x^2}$. Multiplying the DE by this factor gives

$$\frac{1}{5 + x^2} \frac{dy}{dx} - \frac{2x}{(5 + x^2)^2} y = \frac{x}{(5 + x^2)^2}. \tag{1}$$

The left-hand side of the last DE can then be written compactly as the derivative of a product of the integrating factor and $y(x)$:

$$\frac{d}{dx} \left[\frac{1}{5 + x^2} y \right] = \frac{x}{(5 + x^2)^2}. \tag{2}$$

Until you become comfortable with this notion we suggest that you actually multiply the differential equation $dy/dx + P(x)y = f(x)$ by the integrating factor and then verify using the product rule that

$$e^{\int P(x)\,dx} \frac{dy}{dx} + P(x) e^{\int P(x)\,dx} y = e^{\int P(x)\,dx} f(x) \qquad \text{(as in (1))}$$

and

$$\frac{d}{dx} \left[e^{\int P(x)\,dx} y \right] = e^{\int P(x)\,dx} f(x) \qquad \text{(as in (2))}$$

are formally equivalent. After that, once you have computed the integrating factor, you can jump to the last form and proceed with integration. In the case of (2) the integral of the left-hand side is

$$\int \frac{d}{dx} \left[\frac{1}{5 + x^2} y \right] dx = \frac{1}{5 + x^2} y.$$

The integral of the right-hand side is

$$\int \frac{x}{(5 + x^2)^2} \, dx = \frac{1}{2} \int (5 + x^2)^{-2} (2x \, dx) = -\frac{1}{2} (5 + x^2)^{-1} + c.$$

Hence the solution of the DE on $(-\infty, \infty)$ is

$$\frac{1}{5 + x^2} y = -\frac{1}{2} (5 + x^2)^{-1} + c \qquad \text{or} \qquad y = -\frac{1}{2} + c(5 + x^2).$$

Use of Computers   As we have seen in Section 2.2, a CAS can help you find the solution of a DE by performing some of the more complicated integrations required. In addition to this, a CAS can also frequently find directly the solution of a DE. For example, to solve the DE $xy' - xy = e^x$ use

$$\textbf{Dsolve}[\textbf{x y}\,'[\textbf{x}]\textbf{ - x y}[\textbf{x}]\textbf{==Exp}[\textbf{x}], \textbf{y}[\textbf{x}], \textbf{x}] \qquad (\textit{Mathematica})$$

```
dsolve(x*diff(y(x),x)-x*y(x)=exp(x),y(x));      (Maple)
```

To solve an IVP such as $xy' - xy = e^x$, $y(1) = 2$, use

$$\textbf{Dsolve}[\{\textbf{x y}\,'[\textbf{x}]\textbf{ - x y}[\textbf{x}]\textbf{==Exp}[\textbf{x}], \textbf{y}[\textbf{1}]\textbf{==2}\}, \textbf{y}[\textbf{x}], \textbf{x}] \qquad (\textit{Mathematica})$$

```
dsolve({x*diff(y(x),x)-x*y(x)=exp(x),y(1)=2},    (Maple)
    y(x));
```

The output is

$$\{\{y[x] \to e^{x-1}(2 + e\ln x)\}\} \qquad\qquad (\textit{Mathematica})$$

$$y[x] = \left(\ln x + \frac{2}{e}\right)e^x \qquad\qquad (\textit{Maple})$$

In neither case can the output be used subsequently in the CAS to easily obtain numerical values or graphs of the function which is the solution, $y(x)$, of the IVP. The following routines can be used to assign a specific function name to the solution. This function can then be evaluated at a number, differentiated, graphed, etc.

> $\textbf{Clear}[\textbf{y}]$ $(\textit{Mathematica})$
> $\textbf{sol=Flatten}[\textbf{Dsolve}[\{\textbf{x\^{}2 y}\,'[\textbf{x}]\textbf{ - x y}[\textbf{x}]\textbf{==y}[\textbf{x}]\textbf{\^{}2}, \textbf{y}[\textbf{1}]\textbf{==2}\}, \textbf{y}[\textbf{x}], \textbf{x}]];$
> $\textbf{y}[\textbf{x}\_]\textbf{:=Evaluate}[\textbf{y}[\textbf{x}]/\textbf{.sol}]$
> $\textbf{y}[\textbf{x}]$

The semicolon at the end of line of input in *Mathematica* is quite different than in *Maple*, where it is required at the end of every line of input. In *Mathematica* it simply suppresses the display of the output. For example, typing $\textbf{a=2}$ in *Mathematica* will set a equal to 2 and display 2 on the monitor, whereas typing $\textbf{a=2}$; will set a equal to 2, but will not display the 2.

```
sol:=dsolve({x^2*diff(y(x),x)-x*y(x)=y(x)^2,     (Maple)
    y(1)=2}, y(x));
y:=unapply(eval(y(x),sol), x);
```

Special Functions: As discussed in the text, a number of nonelementary functions are defined by integrals. Others, such as the Bessel functions discussed in Chapter 6 of the text, can be defined using infinite series. Virtually all special functions that have been tabulated in standard handbooks are built into *Mathematica* and *Maple*. Below is a table listing all of the special functions used in the text, together with the syntax for each function in both *Mathematica* and *Maple*.

Name	Notation	*Mathematica*	*Maple*
Bessel function of the first kind	$J_n(x)$	**BesselJ[n,x]**	BesselJ(n,x)
Bessel function of the second kind	$Y_n(x)$	**BesselY[n,x]**	BesselY(n,x)
Chebyshev polynomial	$T_n(x)$	**ChebyshevT[n,x]**	ChebyshevT(n,x)
Complementary error function	$\mathrm{erfc}(x)$	**Erfc[x]**	erfc(x)
Dirac delta function	$\delta(t-a)$	**DiracDelta[t-a]**	Dirac(t-a)
Error function	$\mathrm{erf}(x)$	**Erf[x]**	erf(x)
Fresnel sine integral	$S(x)$	**FresnelS[x]**	FresnelS(x)
Gamma	$\Gamma(x)$	**Gamma[x]**	GAMMA(x)
Heaviside (unit step function)	$(t-a)$	**UnitStep[t-a]**	Heaviside(x)
Hermite polynomial	$H_n(x)$	**HermiteH[n,x]**	HermiteH(n,x)
Imaginary error function	$\mathrm{erfi}(x)$	**Erfi[x]**	erfi(x)
Laguerre polynomial	$L_n(x)$	**LaguerreL[n,x]**	orthopoly[L](n,x)
Legendre polynomial	$P_n(x)$	**LegendreP[n,x]**	orthopoly[P](n,x)
Modified Bessel function of the first kind	$I(x)$	**BesselI[n,x]**	BesselI(n,x)
Modified Bessel function of the second kind	$K(x)$	**BesselK[n,x]**	BesselK(n,x)
Sine integral	$Si(x)$	**SinIntegral[x]**	Si(x)

3. For $y' + y = e^{3x}$ an integrating factor is $e^{\int dx} = e^x$ so that $\dfrac{d}{dx}[e^x y] = e^{4x}$ and $y = \frac{1}{4}e^{3x} + ce^{-x}$ for $-\infty < x < \infty$. The transient term is ce^{-x}.

6. For $y' + 2xy = x^3$ an integrating factor is $e^{\int 2x\,dx} = e^{x^2}$ so that $\dfrac{d}{dx}\left[e^{x^2}y\right] = x^3 e^{x^2}$ and $y = \frac{1}{2}x^2 - \frac{1}{2} + ce^{-x^2}$ for $-\infty < x < \infty$. The transient term is ce^{-x^2}.

9. For $y' - \dfrac{1}{x}y = x\sin x$ an integrating factor is $e^{-\int (1/x)dx} = \dfrac{1}{x}$ so that $\dfrac{d}{dx}\left[\dfrac{1}{x}y\right] = \sin x$ and $y = cx - x\cos x$ for $0 < x < \infty$. There is no transient term.

12. For $y' - \dfrac{x}{(1+x)}y = x$ an integrating factor is $e^{-\int [x/(1+x)]dx} = (x+1)e^{-x}$ so that $\dfrac{d}{dx}\left[(x+1)e^{-x}y\right] = x(x+1)e^{-x}$ and $y = -x - \dfrac{2x+3}{x+1} + \dfrac{ce^x}{x+1}$ for $-1 < x < \infty$. There is no transient term.

15. For $\dfrac{dx}{dy} - \dfrac{4}{y}x = 4y^5$ an integrating factor is $e^{-\int (4/y)dy} = e^{\ln y^{-4}} = y^{-4}$ so that $\dfrac{d}{dy}\left[y^{-4}x\right] = 4y$ and $x = 2y^6 + cy^4$ for $0 < y < \infty$. There is no transient term.

18. For $y' + (\cot x)y = \sec^2 x\csc x$ an integrating factor is $e^{\int \cot x\,dx} = e^{\ln|\sin x|} = \sin x$ so that $\dfrac{d}{dx}[(\sin x)\,y] = \sec^2 x$ and $y = \sec x + c\csc x$ for $0 < x < \pi/2$. There is no transient term.

21. For $\dfrac{dr}{d\theta} + r\sec\theta = \cos\theta$ an integrating factor is $e^{\int \sec\theta\,d\theta} = e^{\ln|\sec x + \tan x|} = \sec\theta + \tan\theta$ so that $\dfrac{d}{d\theta}[(\sec\theta + \tan\theta)r] = 1 + \sin\theta$ and $(\sec\theta + \tan\theta)r = \theta - \cos\theta + c$ for $-\pi/2 < \theta < \pi/2$.

24. For $y' + \dfrac{2}{x^2 - 1}y = \dfrac{x+1}{x-1}$ an integrating factor is $e^{\int [2/(x^2-1)]dx} = \dfrac{x-1}{x+1}$ so that $\dfrac{d}{dx}\left[\dfrac{x-1}{x+1}y\right] = 1$ and $(x-1)y = x(x+1) + c(x+1)$ for $-1 < x < 1$.

27. For $y' + \dfrac{1}{x}y = \dfrac{1}{x}e^x$ an integrating factor is $e^{\int (1/x)dx} = x$ so that $\dfrac{d}{dx}[xy] = e^x$ and $y = \dfrac{1}{x}e^x + \dfrac{c}{x}$ for $0 < x < \infty$. If $y(1) = 2$ then $c = 2 - e$ and $y = \dfrac{1}{x}e^x + \dfrac{2-e}{x}$. The solution is defined on $I = (0, \infty)$.

30. For $\dfrac{dT}{dt} - kT = -T_m k$ an integrating factor is $e^{\int (-k)dt} = e^{-kt}$ so that $\dfrac{d}{dt}[e^{-kt}T] = -T_m k e^{-kt}$ and $T = T_m + ce^{kt}$ for $-\infty < t < \infty$. If $T(0) = T_0$ then $c = T_0 - T_m$ and $T = T_m + (T_0 - T_m)e^{kt}$. The solution is defined on $I = (-\infty, \infty)$.

33. For $y' + \dfrac{1}{x+1}\,y = \dfrac{\ln x}{x+1}$ an integrating factor is $e^{\int [1/(x+1)]dx} = x+1$ so that $\dfrac{d}{dx}[(x+1)y] = \ln x$

and

$$y = \frac{x}{x+1}\ln x - \frac{x}{x+1} + \frac{c}{x+1}$$

for $0 < x < \infty$. If $y(1) = 10$ then $c = 21$ and $y = \dfrac{x}{x+1}\ln x - \dfrac{x}{x+1} + \dfrac{21}{x+1}$. The solution is

defined on $I = (0,\,\infty)$.

36. For $y' + (\tan x)y = \cos^2 x$ an integrating factor is $e^{\int \tan x\,dx} = e^{\ln|\sec x|} = \sec x$ so that $\dfrac{d}{dx}[(\sec x)\,y] = \cos x$ and $y = \sin x\cos x + c\cos x$ for $-\pi/2 < x < \pi/2$. If $y(0) = -1$ then $c = -1$ and $y = \sin x\cos x - \cos x$. The solution is defined on $I = (-\pi/2,\,\pi/2)$.

39. For $y' + 2xy = f(x)$ an integrating factor is e^{x^2} so that

$$ye^{x^2} = \begin{cases} \frac{1}{2}e^{x^2} + c_1, & 0 \le x < 1 \\ c_2, & x \ge 1. \end{cases}$$

If $y(0) = 2$ then $c_1 = 3/2$ and for continuity we must have $c_2 = \frac{1}{2}e + \frac{3}{2}$ so that

$$y = \begin{cases} \frac{1}{2} + \frac{3}{2}e^{-x^2}, & 0 \le x < 1 \\ \left(\frac{1}{2}e + \frac{3}{2}\right)e^{-x^2}, & x \ge 1. \end{cases}$$

42. For $y' + e^x y = 1$ an integrating factor is e^{e^x}. Thus

$$\frac{d}{dx}\left[e^{e^x}y\right] = e^{e^x} \quad \text{and} \quad e^{e^x}y = \int_0^x e^{e^t}\,dt + c.$$

From $y(0) = 1$ we get $c = e$, so $y = e^{-e^x}\int_0^x e^{e^t}\,dt + e^{1-e^x}$.

When $y' + e^x y = 0$ we can separate variables and integrate:

$$\frac{dy}{y} = -e^x\,dx \quad \text{and} \quad \ln|y| = -e^x + c.$$

Thus $y = c_1 e^{-e^x}$. From $y(0) = 1$ we get $c_1 = e$, so $y = e^{1-e^x}$.

When $y' + e^x y = e^x$ we can see by inspection that $y = 1$ is a solution.

2.4 | Exact Equations

The terminology and concepts listed below provide an outline of the main ideas encountered in this section. These can be useful when preparing for a quiz or test.

Terminology and Concepts

- differential of a function f of two variables

- exact DE

- criterion for exactness of $M(x, y)\, dx + N(x, y)\, dy = 0$

- use of an integrating factor to change a nonexact DE into one that is exact

The basic skills listed below summarize the more mechanical types of problems encountered in the exercise set for this section.

Basic Skills

- determine whether a DE is exact

- solve an exact DE

- find and use an integrating factor to make a DE exact

Partial Differentiation   In this section you will have to use **partial differentiation** and a reverse process which could be called **partial integration** or **partial antidifferentiation.**

Suppose $f(x, y)$ is a function of two variables defined in a region R of the plane. The mechanics of computing the partial derivative of f with respect to x, written $\partial f / \partial x$, can be summarized simply as:

> *To compute $\dfrac{\partial f}{\partial x}$, use the rules of ordinary differentiation with respect to x while treating y as a constant.*

For example, if $f(x, y) = 3x^4 y^2 + e^{5y} + 4\cos xy$, then by treating y as a constant we obtain

$$\frac{\partial f}{\partial x} = y^2 \frac{\partial}{\partial x} 3x^4 + \frac{\partial}{\partial x} e^{5y} + 4 \frac{\partial}{\partial x} \cos xy^2 = y^2 \cdot 12x^3 + 0 + 4(-y^2 \sin xy^2)$$

$$= 12x^3 y^2 - 4y^2 \sin xy^2.$$

On the other hand, to compute the partial derivative of f with respect to y, written $\partial f/\partial y$, use the rules of ordinary differentiation, except treat x as a constant. For example, for the function f in the preceding example, we have

$$\frac{\partial f}{\partial y} = 3x^4 \frac{\partial}{\partial y} y^2 + \frac{\partial}{\partial y} e^{5y} + 4\frac{\partial}{\partial y} \cos xy^2 = 3x^4 \cdot 2y + 5e^{5y} + 4(-x \cdot 2y \sin xy^2)$$

$$= 6x^4 y + 5e^{5y} - 8xy \sin xy^2.$$

Now suppose that we are given the partial derivative of a function f, say $\partial f/\partial x = M(x, y)$. We can recover f up to an "additive constant" by partial integration with respect to x:

$$f(x, y) = \int M(x, y)\, dx.$$

In the foregoing integral we use the rules of ordinary indefinite integration with respect to x while treating y as a constant. If $\partial f/\partial y = N(x, y)$, then by partial integration with respect to y we have

$$f(x, y) = \int N(x, y)\, dy,$$

where x is treated as a constant. For example, if $\partial f/\partial x = x + y$, then

$$f(x, y) = \int (x + y)\, dx = \frac{1}{2}x^2 + xy + g(y).$$

Here any function $g(y)$ is a "constant" of integration since

$$\frac{\partial f}{\partial x} = \frac{1}{2}\frac{\partial}{\partial x} x^2 + y\frac{\partial}{\partial x} x + \frac{\partial}{\partial x} g(y) = \frac{1}{2} \cdot 2x + y \cdot 1 + 0 = x + y.$$

As another example, suppose that $\partial f/\partial y = 10e^{-x^2 y}$, then, with x held constant,

$$f(x, y) = \int 10e^{-x^2 y}\, dy = \frac{10}{-x^2} \int e^{-x^2 y}(-x^2\, dy) = -\frac{10}{x^2} e^{-x^2 y} + h(x),$$

where this time the "constant" of integration is $h(x)$. You are encouraged to verify the last result by partial differentiation with respect to y.

Use of Computers   To obtain the partial derivative of $f(x, y) = xe^y + x^2 y^3$ with respect to y in a CAS use

Clear[f]   (*Mathematica*)

f[x_, y_]:= x Exp[y] + x^2 y^3

f[x, y]

D[f[x, y], y]

```
f:=(x,y)->x*exp(y)+x^2*y^3;        (Maple)
diff(f(x,y),y);      or      D[2](f);
```

In *Maple*, `diff(f(x,y),y);` returns the expression $xe^y + 3x^2y^2$, which cannot be directly evaluated at a point. On the other hand, `D[2](f);` returns the function $(x, y) \rightarrow xe^y + 3x^2y^2$. In this case, the 2 in `D[2]` signifies the partial derivative with respect to the second variable, y, in the definition of f.

3. Let $M = 5x + 4y$ and $N = 4x - 8y^3$ so that $M_y = 4 = N_x$. From $f_x = 5x + 4y$ we obtain $f = \frac{5}{2}x^2 + 4xy + h(y)$, $h'(y) = -8y^3$, and $h(y) = -2y^4$. A solution is $\frac{5}{2}x^2 + 4xy - 2y^4 = c$.

6. Let $M = 4x^3 - 3y \sin 3x - y/x^2$ and $N = 2y - 1/x + \cos 3x$ so that $M_y = -3 \sin 3x - 1/x^2$ and $N_x = 1/x^2 - 3 \sin 3x$. The equation is not exact.

9. Let $M = y^3 - y^2 \sin x - x$ and $N = 3xy^2 + 2y \cos x$ so that $M_y = 3y^2 - 2y \sin x = N_x$. From $f_x = y^3 - y^2 \sin x - x$ we obtain $f = xy^3 + y^2 \cos x - \frac{1}{2}x^2 + h(y)$, $h'(y) = 0$, and $h(y) = 0$. A solution is $xy^3 + y^2 \cos x - \frac{1}{2}x^2 = c$.

12. Let $M = 3x^2y + e^y$ and $N = x^3 + xe^y - 2y$ so that $M_y = 3x^2 + e^y = N_x$. From $f_x = 3x^2y + e^y$ we obtain $f = x^3y + xe^y + h(y)$, $h'(y) = -2y$, and $h(y) = -y^2$. A solution is $x^3y + xe^y - y^2 = c$.

15. Let $M = x^2y^3 - 1/(1 + 9x^2)$ and $N = x^3y^2$ so that $M_y = 3x^2y^2 = N_x$. From $f_x = x^2y^3 - 1/(1 + 9x^2)$ we obtain $f = \frac{1}{3}x^3y^3 - \frac{1}{3}\arctan(3x) + h(y)$, $h'(y) = 0$, and $h(y) = 0$. A solution is $x^3y^3 - \arctan(3x) = c$.

18. Let $M = 2y \sin x \cos x - y + 2y^2e^{xy^2}$ and $N = -x + \sin^2 x + 4xye^{xy^2}$ so that

$$M_y = 2 \sin x \cos x - 1 + 4xy^3e^{xy^2} + 4ye^{xy^2} = N_x.$$

From $f_x = 2y \sin x \cos x - y + 2y^2e^{xy^2}$ we obtain $f = y \sin^2 x - xy + 2e^{xy^2} + h(y)$, $h'(y) = 0$, and $h(y) = 0$. A solution is $y \sin^2 x - xy + 2e^{xy^2} = c$.

21. Let $M = x^2 + 2xy + y^2$ and $N = 2xy + x^2 - 1$ so that $M_y = 2(x + y) = N_x$. From $f_x = x^2 + 2xy + y^2$ we obtain $f = \frac{1}{3}x^3 + x^2y + xy^2 + h(y)$, $h'(y) = -1$, and $h(y) = -y$. The solution is $\frac{1}{3}x^3 + x^2y + xy^2 - y = c$. If $y(1) = 1$ then $c = 4/3$ and a solution of the initial-value problem is $\frac{1}{3}x^3 + x^2y + xy^2 - y = \frac{4}{3}$.

24. Let $M = t/2y^4$ and $N = (3y^2 - t^2)/y^5$ so that $M_y = -2t/y^5 = N_t$. From $f_t = t/2y^4$ we obtain $f = \dfrac{t^2}{4y^4} + h(y)$, $h'(y) = \dfrac{3}{y^3}$, and $h(y) = -\dfrac{3}{2y^2}$. The solution is $\dfrac{t^2}{4y^4} - \dfrac{3}{2y^2} = c$. If $y(1) = 1$ then $c = -5/4$ and a solution of the initial-value problem is $\dfrac{t^2}{4y^4} - \dfrac{3}{2y^2} = -\dfrac{5}{4}$.

27. Equating $M_y = 3y^2 + 4kxy^3$ and $N_x = 3y^2 + 40xy^3$ we obtain $k = 10$.

30. Let $M = \left(x^2 + 2xy - y^2\right) / \left(x^2 + 2xy + y^2\right)$ and $N = \left(y^2 + 2xy - x^2\right) / \left(y^2 + 2xy + x^2\right)$ so that $M_y = -4xy/(x+y)^3 = N_x$. From $f_x = \left(x^2 + 2xy + y^2 - 2y^2\right)/(x+y)^2$ we obtain

$$f = x + \frac{2y^2}{x+y} + h(y), \ h'(y) = -1, \text{ and } h(y) = -y. \text{ A solution of the differential equation is}$$
$x^2 + y^2 = c(x+y)$.

33. We note that $(N_x - M_y)/M = 2/y$, so an integrating factor is $e^{\int 2dy/y} = y^2$. Let $M = 6xy^3$ and $N = 4y^3 + 9x^2y^2$ so that $M_y = 18xy^2 = N_x$. From $f_x = 6xy^3$ we obtain $f = 3x^2y^3 + h(y)$, $h'(y) = 4y^3$, and $h(y) = y^4$. A solution of the differential equation is $3x^2y^3 + y^4 = c$.

36. We note that $(N_x - M_y)/M = -3/y$, so an integrating factor is $e^{-3\int dy/y} = 1/y^3$. Let

$$M = (y^2 + xy^3)/y^3 = 1/y + x \quad \text{and} \quad N = (5y^2 - xy + y^3 \sin y)/y^3 = 5/y - x/y^2 + \sin y,$$

so that $M_y = -1/y^2 = N_x$. From $f_x = 1/y + x$ we obtain $f = x/y + \frac{1}{2}x^2 + h(y)$, $h'(y) = 5/y + \sin y$, and $h(y) = 5 \ln |y| - \cos y$. A solution of the differential equation is $x/y + \frac{1}{2}x^2 + 5 \ln |y| - \cos y = c$.

39. (a) Implicitly differentiating $x^3 + 2x^2y + y^2 = c$ and solving for dy/dx we obtain

$$3x^2 + 2x^2 \frac{dy}{dx} + 4xy + 2y \frac{dy}{dx} = 0 \quad \text{and} \quad \frac{dy}{dx} = -\frac{3x^2 + 4xy}{2x^2 + 2y}.$$

By writing the last equation in differential form we get $(4xy + 3x^2)dx + (2y + 2x^2)dy = 0$.

(b) Setting $x = 0$ and $y = -2$ in $x^3 + 2x^2y + y^2 = c$ we find $c = 4$, and setting $x = y = 1$ we also find $c = 4$. Thus, both initial conditions determine the same implicit solution.

(c) Solving $x^3 + 2x^2y + y^2 = 4$ for y we get

$$y_1(x) = -x^2 - \sqrt{4 - x^3 + x^4}$$

and

$$y_2(x) = -x^2 + \sqrt{4 - x^3 + x^4}.$$

Observe in the figure that $y_1(0) = -2$ and $y_2(1) = 1$.

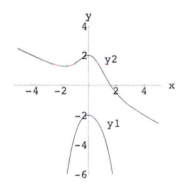

2.5 | Solutions by Substitutions

The terminology and concepts listed below provide an outline of the main ideas encountered in this section. These can be useful when preparing for a quiz or test.

Terminology and Concepts

- homogeneous DE

- Bernoulli's equation

- use of a substitution to reduce a DE of the form $dy/dx = f(Ax + By + C)$ to a separable DE

The basic skills listed below summarize the more mechanical types of problems encountered in the exercise set for this section.

Basic Skills

- recognize and solve a homogeneous DE using an appropriate substitution

- recognize a Bernoulli equation and solve it by reducing it to a linear first-order DE

- solve a DE of the form $dy/dx = f(Ax + By + C)$ using an appropriate substitution

Homogeneous DEs   As mentioned in the text, either substitution $y = ux$ or $x = vy$ will reduce a homogeneous DE $M(x, y)\,dx + N(x, y)\,dy = 0$ to one with separable variables. As a rule of thumb, try the substitution $y = ux$ whenever the coefficient N is simpler than M. On the other hand, try the substitution $x = vy$ whenever the coefficient M is simpler than N.

3. Letting $x = vy$ we have

$$vy(v\,dy + y\,dv) + (y - 2vy)\,dy = 0$$

$$vy^2\,dv + y\left(v^2 - 2v + 1\right)dy = 0$$

$$\frac{v\,dv}{(v-1)^2} + \frac{dy}{y} = 0$$

$$\ln|v - 1| - \frac{1}{v - 1} + \ln|y| = c$$

$$\ln\left|\frac{x}{y} - 1\right| - \frac{1}{x/y - 1} + \ln y = c$$

$$(x - y)\ln|x - y| - y = c(x - y).$$

6. Letting $y = ux$ and using partial fractions, we have

$$\left(u^2 x^2 + ux^2\right) dx + x^2 (u\,dx + x\,du) = 0$$

$$x^2 \left(u^2 + 2u\right) dx + x^3\,du = 0$$

$$\frac{dx}{x} + \frac{du}{u(u+2)} = 0$$

$$\ln|x| + \frac{1}{2}\ln|u| - \frac{1}{2}\ln|u+2| = c$$

$$\frac{x^2 u}{u+2} = c_1$$

$$x^2 \frac{y}{x} = c_1\left(\frac{y}{x} + 2\right)$$

$$x^2 y = c_1(y + 2x).$$

9. Letting $y = ux$ we have

$$-ux\,dx + (x + \sqrt{u}\,x)(u\,dx + x\,du) = 0$$

$$(x^2 + x^2\sqrt{u}\,)\,du + xu^{3/2}\,dx = 0$$

$$\left(u^{-3/2} + \frac{1}{u}\right) du + \frac{dx}{x} = 0$$

$$-2u^{-1/2} + \ln|u| + \ln|x| = c$$

$$\ln|y/x| + \ln|x| = 2\sqrt{x/y} + c$$

$$y(\ln|y| - c)^2 = 4x.$$

12. Letting $y = ux$ we have

$$(x^2 + 2u^2 x^2)dx - ux^2(u\,dx + x\,du) = 0$$

$$x^2(1 + u^2)dx - ux^3\,du = 0$$

$$\frac{dx}{x} - \frac{u\,du}{1 + u^2} = 0$$

$$\ln|x| - \frac{1}{2}\ln(1 + u^2) = c$$

$$\frac{x^2}{1 + u^2} = c_1$$

$$x^4 = c_1(x^2 + y^2).$$

Using $y(-1) = 1$ we find $c_1 = 1/2$. The solution of the initial-value problem is $2x^4 = y^2 + x^2$.

15. From $y' + \frac{1}{x}y = \frac{1}{x}y^{-2}$ and $w = y^3$ we obtain $\frac{dw}{dx} + \frac{3}{x}w = \frac{3}{x}$. An integrating factor is x^3 so that $x^3w = x^3 + c$ or $y^3 = 1 + cx^{-3}$.

18. From $y' - \left(1 + \frac{1}{x}\right)y = y^2$ and $w = y^{-1}$ we obtain $\frac{dw}{dx} + \left(1 + \frac{1}{x}\right)w = -1$. An integrating factor is xe^x so that $xe^x w = -xe^x + e^x + c$ or $y^{-1} = -1 + \frac{1}{x} + \frac{c}{x}e^{-x}$.

21. From $y' - \frac{2}{x}y = \frac{3}{x^2}y^4$ and $w = y^{-3}$ we obtain $\frac{dw}{dx} + \frac{6}{x}w = -\frac{9}{x^2}$. An integrating factor is x^6 so that $x^6w = -\frac{9}{5}x^5 + c$ or $y^{-3} = -\frac{9}{5}x^{-1} + cx^{-6}$. If $y(1) = \frac{1}{2}$ then $c = \frac{49}{5}$ and $y^{-3} = -\frac{9}{5}x^{-1} + \frac{49}{5}x^{-6}$.

24. Let $u = x + y$ so that $du/dx = 1 + dy/dx$. Then $\frac{du}{dx} - 1 = \frac{1-u}{u}$ or $u\,du = dx$. Thus $\frac{1}{2}u^2 = x + c$ or $u^2 = 2x + c_1$, and $(x + y)^2 = 2x + c_1$.

27. Let $u = y - 2x + 3$ so that $du/dx = dy/dx - 2$. Then $\frac{du}{dx} + 2 = 2 + \sqrt{u}$ or $\frac{1}{\sqrt{u}}du = dx$. Thus $2\sqrt{u} = x + c$ and $2\sqrt{y - 2x + 3} = x + c$.

30. Let $u = 3x + 2y$ so that $du/dx = 3 + 2\,dy/dx$. Then $\frac{du}{dx} = 3 + \frac{2u}{u+2} = \frac{5u+6}{u+2}$ and $\frac{u+2}{5u+6}du = dx$. Now by long division

$$\frac{u+2}{5u+6} = \frac{1}{5} + \frac{4}{25u+30}$$

so we have

$$\int \left(\frac{1}{5} + \frac{4}{25u+30}\right)du = dx$$

and $\frac{1}{5}u + \frac{4}{25}\ln|25u + 30| = x + c$. Thus

$$\frac{1}{5}(3x + 2y) + \frac{4}{25}\ln|75x + 50y + 30| = x + c.$$

Setting $x = -1$ and $y = -1$ we obtain $c = \frac{4}{25} \ln 95$. The solution is

$$\frac{1}{5}(3x + 2y) + \frac{4}{25} \ln|75x + 50y + 30| = x + \frac{4}{25} \ln 95$$

or $5y - 5x + 2\ln|75x + 50y + 30| = 2\ln 95$.

2.6 A Numerical Method

The terminology and concepts listed below provide an outline of the main ideas encountered in this section. These can be useful when preparing for a quiz or test.

Terminology and Concepts

- direction field

- linearization of a function at a point

- Euler's method

- absolute error

- percentage relative error

- numerical solver

The basic skills listed below summarize the more mechanical types of problems encountered in the exercise set for this section.

Basic Skills

- use Euler's method to approximate the solution of a first-order DE

Use of Computers

Both *Mathematica* and *Maple* contain sophisticated programming languages. Below are examples of how each CAS can be used to implement Euler's method and display a table similar to the ones shown in Example 2 in this section of the text. Here we will consider the IVP $y' = f(x, y) = 0.2xy$, $y(1) = 1$, and use Euler's method to approximate $y(1.5)$ using the step size $h = 0.1$. To compute the absolute error at each step we will use the fact that the analytic solution is $y = g(x) = e^{0.1(x^2-1)}$.

```
Clear[f, g, u, v, t, err]                          (Mathematica)
f[x_, y_]:= 0.2x y        (* Input line *)
Print["y'= ", f[x, y]]
g[x_]:= Exp[0.1(x^2 - 1)]        (* Input line *)
Print["y(x) = ", g[x]]
h = 0.1; a = 1; b = 1.5; y0 = 1;        (* Input line *)
```

```
u[0] = a; v[0] = y0;
u[n_]:= u[n] = u[n - 1] + h
v[n_]:= v[n] = v[n - 1] + h f[u[n - 1], v[n - 1]]
t[n_]:= t[n] = g[u[n]]
err[n_]:= err[n] = Abs[t[n] - v[n]]
approxtab = Table[{u[n], v[n], t[n], err[n]}, {n, 0, Round[(b - a)/h]}];
tableheader = {"x", "y", "actual value", "abs. error"};
Prepend[approxtab, tableheader]//TableForm
```

The use of the form $u[n_]:= u[n] = u[n - 1] + h$ in the *Mathematica* code above is a way to define a function that remembers a value once it has been computed. This greatly speeds up the running of a program that involves a recursively defined function. To make it easier to reuse the routine for subsequent problems, the lines containing inputs to the routine can be commented, as shown above, or color coded.

```
f:=(x,y)->0.2*x*y;          #Input line              (Maple)
g:=x->exp(0.1*(x^2-1));        #Input line
h:=0.1; a:=1; b:=1.5; y0:=1;        #Input line
u:=n->u(n-1)+h; u(0):=a;
v:=n->v(n-1)+h*f(u(n-1),v(n-1)); v(0):=y0;
t:=n->g(u(n));
err:=n->abs(t(n)-v(n));
u(-1):='x'; v(-1):='y'; t(-1):='actual_value';
err(-1):='abs._error';
with(linalg)
matrix(2+round((b-a)/h),4,(i,j)->piecewise(j=1,u(i-2),
    j=2,v(i-2),j=3,t(i-2),j=4,err(i-2)));
```

3. Separating variables and integrating, we have

$$\frac{dy}{y} = dx \qquad \text{and} \qquad \ln |y| = x + c.$$

Thus $y = c_1 e^x$ and, using $y(0) = 1$, we find $c = 1$, so $y = e^x$ is the solution of the initial-value problem.

$h=0.1$

x_n	y_n	Actual Value	Abs. Error	% Rel. Error
0.00	1.0000	1.0000	0.0000	0.00
0.10	1.1000	1.1052	0.0052	0.47
0.20	1.2100	1.2214	0.0114	0.93
0.30	1.3310	1.3499	0.0189	1.40
0.40	1.4641	1.4918	0.0277	1.86
0.50	1.6105	1.6487	0.0382	2.32
0.60	1.7716	1.8221	0.0506	2.77
0.70	1.9487	2.0138	0.0650	3.23
0.80	2.1436	2.2255	0.0820	3.68
0.90	2.3579	2.4596	0.1017	4.13
1.00	2.5937	2.7183	0.1245	4.58

$h=0.05$

x_n	y_n	Actual Value	Abs. Error	% Rel. Error
0.00	1.0000	1.0000	0.0000	0.00
0.05	1.0500	1.0513	0.0013	0.12
0.10	1.1025	1.1052	0.0027	0.24
0.15	1.1576	1.1618	0.0042	0.36
0.20	1.2155	1.2214	0.0059	0.48
0.25	1.2763	1.2840	0.0077	0.60
0.30	1.3401	1.3499	0.0098	0.72
0.35	1.4071	1.4191	0.0120	0.84
0.40	1.4775	1.4918	0.0144	0.96
0.45	1.5513	1.5683	0.0170	1.08
0.50	1.6289	1.6487	0.0198	1.20
0.55	1.7103	1.7333	0.0229	1.32
0.60	1.7959	1.8221	0.0263	1.44
0.65	1.8856	1.9155	0.0299	1.56
0.70	1.9799	2.0138	0.0338	1.68
0.75	2.0789	2.1170	0.0381	1.80
0.80	2.1829	2.2255	0.0427	1.92
0.85	2.2920	2.3396	0.0476	2.04
0.90	2.4066	2.4596	0.0530	2.15
0.95	2.5270	2.5857	0.0588	2.27
1.00	2.6533	2.7183	0.0650	2.39

6. $h=0.1$

x_n	y_n
0.00	1.0000
0.10	1.1000
0.20	1.2220
0.30	1.3753
0.40	1.5735
0.50	1.8371

$h=0.05$

x_n	y_n
0.00	1.0000
0.05	1.0500
0.10	1.1053
0.15	1.1668
0.20	1.2360
0.25	1.3144
0.30	1.4039
0.35	1.5070
0.40	1.6267
0.45	1.7670
0.50	1.9332

9. $h=0.1$

x_n	y_n
1.00	1.0000
1.10	1.0000
1.20	1.0191
1.30	1.0588
1.40	1.1231
1.50	1.2194

$h=0.05$

x_n	y_n
1.00	1.0000
1.05	1.0000
1.10	1.0049
1.15	1.0147
1.20	1.0298
1.25	1.0506
1.30	1.0775
1.35	1.1115
1.40	1.1538
1.45	1.2057
1.50	1.2696

12. Tables of values were computed using the Euler and RK4 methods. The resulting points were plotted and joined using **ListPlot** in *Mathematica*.

2.R   Chapter 2 in Review

We see that when $x = 0$, $y = 0$, so the initial-value problem has an infinite number of solutions for $k = 0$ and no solutions for $k \neq 0$.

3. True; $y = k_2/k_1$ is always a solution for $k_1 \neq 0$.

6. False, because $r\theta + r + \theta + 1 = (r + 1)(\theta + 1)$ and the differential equation can be written as

$$\frac{dr}{r + 1} = (\theta + 1)d\theta.$$

9. The differential equation is separable so

$$\frac{dy}{y} = e^x dx \qquad \text{implies} \qquad \ln|y| = e^{e^x} + c,$$

and thus $y = c_1 e^{e^x}$ is the general solution of the differential equation.

12. An example of an autonomous linear first-order differential equation with a single critical point at -3 is $\dfrac{dy}{dx} = y + 3$, whereas an autonomous nonlinear first-order differential equation with a single critical point -3 is $\dfrac{dy}{dx} = (y + 3)^2$.

15. When n is odd, $x^n < 0$ for $x < 0$ and $x^n > 0$ for $x > 0$. In this case 0 is unstable. When n is even, $x^n > 0$ for $x < 0$ and for $x > 0$. In this case 0 is semi-stable. When n is odd, $-x^n > 0$ for $x < 0$ and $-x^n < 0$ for $x > 0$. In this case 0 is asymptotically stable. When n is even, $-x^n < 0$ for $x < 0$ and for $x > 0$. In this case 0 is semi-stable. Technically, 0^0 is an indeterminant form; however for all values of x except 0, $x^0 = 1$. Thus, we define 0^0 to be 1 in this case.

18. **(a)** linear in y, homogeneous, exact

(b) linear in x

(c) separable, exact, linear in x and y

(d) Bernoulli in x

(e) separable

(f) separable, linear in x, Bernoulli

(g) linear in x

(h) homogeneous

(i) Bernoulli

(j) homogeneous, exact, Bernoulli

(k) linear in x and y, exact, separable, homogeneous

(l) exact, linear in y

(m) homogeneous

(n) separable

21. The differential equation

$$\frac{dy}{dx} + \frac{2}{6x+1}y = -\frac{3x^2}{6x+1}y^{-2}$$

is Bernoulli. Using $w = y^3$, we obtain the linear equation

$$\frac{dw}{dx} + \frac{6}{6x+1}w = -\frac{9x^2}{6x+1}.$$

An integrating factor is $6x + 1$, so

$$\frac{d}{dx}[(6x+1)w] = -9x^2,$$

$$w = -\frac{3x^3}{6x+1} + \frac{c}{6x+1},$$

and

$$(6x+1)y^3 = -3x^3 + c.$$

Note: The differential equation is also exact.

24. Letting $u = 2x + y + 1$ we have

$$\frac{du}{dx} = 2 + \frac{dy}{dx},$$

and so the given differential equation is transformed into

$$u\left(\frac{du}{dx} - 2\right) = 1 \quad \text{or} \quad \frac{du}{dx} = \frac{2u + 1}{u}.$$

Separating variables and integrating we get

$$\frac{u}{2u + 1}\, du = dx$$

$$\left(\frac{1}{2} - \frac{1}{2}\frac{1}{2u + 1}\right) du = dx$$

$$\frac{1}{2}u - \frac{1}{4}\ln|2u + 1| = x + c$$

$$2u - \ln|2u + 1| = 2x + c_1.$$

Resubstituting for u gives the solution

$$4x + 2y + 2 - \ln|4x + 2y + 3| = 2x + c_1$$

or

$$2x + 2y + 2 - \ln|4x + 2y + 3| = c_1.$$

27. The differential equation has the form $(d/dx)\,[(\sin x)y] = 0$. Integrating, we have $(\sin x)y = c$ or $y = c/\sin x$. The initial condition implies $c = -2\sin(7\pi/6) = 1$. Thus, $y = 1/\sin x = \csc x$, where the interval $\pi < x < 2\pi$ is chosen to include $x = 7\pi/6$.

30. (a) The differential equation is homogeneous and we let $y = ux$. Then

$$(x^2 - y^2)\,dx + xy\,dy = 0$$

$$(x^2 - u^2x^2)\,dx + ux^2(u\,dx + x\,du) = 0$$

$$dx + ux\,du = 0$$

$$u\,du = -\frac{dx}{x}$$

$$\frac{1}{2}u^2 = -\ln|x| + c$$

$$\frac{y^2}{x^2} = -2\ln|x| + c_1.$$

The initial condition gives $c_1 = 2$, so an implicit solution is $y^2 = x^2(2 - 2\ln|x|)$.

(b) Solving for y in part (a) and being sure that the initial condition is still satisfied, we have $y = -\sqrt{2}\,|x|(1 - \ln|x|)^{1/2}$, where $-e \le x \le e$ so that $1 - \ln|x| \ge 0$. The graph of this function indicates that the derivative is not defined at $x = 0$ and $x = e$. Thus, the solution of the initial-value problem is $y = -\sqrt{2}\,x(1 - \ln x)^{1/2}$, for $0 < x < e$.

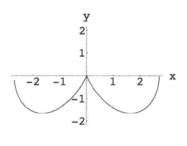

33. Since the differential equation is autonomous, all lineal elements on a given horizontal line have the same slope. The direction field is then as shown in the figure at the right. It appears from the figure that the differential equation has critical points at -2 (an attractor) and at 2 (a repeller). Thus, -2 is an asymptotically stable critical point and 2 is an unstable critical point.

3 MODELING WITH FIRST-ORDER DIFFERENTIAL EQUATIONS

| 3.1 | Linear Models |

The terminology and concepts listed below provide an outline of the main ideas encountered in this section. These can be useful when preparing for a quiz or test.

Terminology and Concepts

- growth and decay
- growth and decay constant
- half-life
- carbon dating
- Newton's law of cooling/warming
- mixtures
- series circuits
- steady-state part of a solution
- transient term in solution

The basic skills listed below summarize the more mechanical types of problems encountered in the exercise set for this section.

Basic Skills

- set up and solve DEs modeling populations that exhibit exponential growth or decay
- set up and solve DEs involving Newton's law of cooling/warming
- set up and solve DEs modeling mixtures of fluids
- set up and solve DEs modeling series circuits
- set up and solve DEs modeling falling objects both with and without air resistance

Half-life To drive the point home that "the usual carbon-14 technique is limited to about nine half-lives of the isotope," we will estimate the amount of C-14 that would remain in a piece of wood from a tree about 50,000 years after the tree is cut down. We use the fact that the half-life of C-14 is approximately 5,600 years. Then, in $2(5,600) = 12,200$ years, $\frac{1}{4}$ of the original amount of C-14 remains; in $3(5,600) = 16,800$ years, $\frac{1}{8}$ of the original amount remains. Continuing in this fashion, in $9(5,600) = 50,400$ years, $\frac{1}{512}$ (or only about 0.2%) of the original amount of C-14 remains.

Growth and Decay

3. Let $P = P(t)$ be the population at time t. Then $dP/dt = kP$ and $P = ce^{kt}$. From $P(0) = c = 500$ we see that $P = 500e^{kt}$. Since 15% of 500 is 75, we have $P(10) = 500e^{10k} = 575$. Solving for k, we get $k = \frac{1}{10}\ln\frac{575}{500} = \frac{1}{10}\ln 1.15$. When $t = 30$,

$$P(30) = 500e^{(1/10)(\ln 1.15)30} = 500e^{3\ln 1.15} = 760 \text{ years}$$

and

$$P'(30) = kP(30) = \frac{1}{10}(\ln 1.15)760 = 10.62 \text{ persons/year.}$$

6. Let $A = A(t)$ be the amount present at time t. From $dA/dt = kA$ and $A(0) = 100$ we obtain $A = 100e^{kt}$. Using $A(6) = 97$ we find $k = \frac{1}{6}\ln 0.97$. Then $A(24) = 100e^{(1/6)(\ln 0.97)24} = 100(0.97)^4 \approx 88.5$ mg.

9. Let $I = I(t)$ be the intensity, t the thickness, and $I(0) = I_0$. If $dI/dt = kI$ and $I(3) = 0.25I_0$, then $I = I_0e^{kt}$, $k = \frac{1}{3}\ln 0.25$, and $I(15) = 0.00098I_0$.

Carbon Dating

12. From Example 3 in the text, the amount of carbon present at time t is $A(t) = A_0e^{-0.00012378t}$. Letting $t = 660$ and solving for A_0 we have $A(660) = A_0e^{-0.0001237(660)} = 0.921553A_0$. Thus, approximately 92% of the original amount of C-14 remained in the cloth as of 1988.

Newton's Law of Cooling/Warming

15. We use the fact that the boiling temperature for water is 100° C. Now assume that $dT/dt = k(T - 100)$ so that $T = 100 + ce^{kt}$. If $T(0) = 20°$ and $T(1) = 22°$, then $c = -80$ and $k = \ln(39/40) \approx -0.0253$. Then $T(t) = 100 - 80e^{-0.0253t}$, and when $T = 90$, $t = 82.1$ seconds. If $T(t) = 98°$ then $t = 145.7$ seconds.

18. (a) The initial temperature of the bath is $T_m(0) = 60°$, so in the short term the temperature of the chemical, which starts at $80°$, should decrease or cool. Over time, the temperature of the bath will increase toward $100°$ since $e^{-0.1t}$ decreases from 1 toward 0 as t increases from 0. Thus, in the long term, the temperature of the chemical should increase or warm toward $100°$.

(b) Adapting the model for Newton's law of cooling, we have

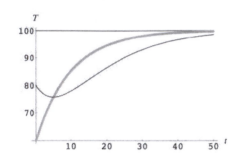

$$\frac{dT}{dt} = -0.1(T - 100 + 40e^{-0.1t}), \quad T(0) = 80.$$

Writing the differential equation in the form

$$\frac{dT}{dt} + 0.1T = 10 - 4e^{-0.1t}$$

we see that it is linear with integrating factor $e^{\int 0.1\,dt} = e^{0.1t}$. Thus

$$\frac{d}{dt}[e^{0.1t}T] = 10e^{0.1t} - 4$$

$$e^{0.1t}T = 100e^{0.1t} - 4t + c$$

and

$$T(t) = 100 - 4te^{-0.1t} + ce^{-0.1t}.$$

Now $T(0) = 80$ so $100 + c = 80$, $c = -20$ and

$$T(t) = 100 - 4te^{-0.1t} - 20e^{-0.1t} = 100 - (4t + 20)e^{-0.1t}.$$

The thinner curve verifies the prediction of cooling followed by warming toward $100°$. The wider curve shows the temperature T_m of the liquid bath.

Mixtures

21. From $dA/dt = 4 - A/50$ we obtain $A = 200 + ce^{-t/50}$. If $A(0) = 30$ then $c = -170$ and $A = 200 - 170e^{-t/50}$.

24. From Problem 23 the number of pounds of salt in the tank at time t is $A(t) = 1000 - 1000e^{-t/100}$. The concentration at time t is $c(t) = A(t)/500 = 2 - 2e^{-t/100}$. Therefore $c(5) = 2 - 2e^{-1/20} = 0.0975\,\text{lb/gal}$ and $\lim_{t\to\infty} c(t) = 2$. Solving $c(t) = 1 = 2 - 2e^{-t/100}$ for t we obtain $t = 100\ln 2 \approx 69.3\,\text{min}$.

27. From

$$\frac{dA}{dt} = 3 - \frac{4A}{100 + (6-4)t} = 3 - \frac{2A}{50+t}$$

we obtain $A = 50 + t + c(50+t)^{-2}$. If $A(0) = 10$ then $c = -100,000$ and $A(30) = 64.38$ pounds.

Series Circuits

30. Assume $L\, di/dt + Ri = E(t)$, $E(t) = E_0 \sin \omega t$, and $i(0) = i_0$ so that

$$i = \frac{E_0 R}{L^2 \omega^2 + R^2} \sin \omega t - \frac{E_0 L \omega}{L^2 \omega^2 + R^2} \cos \omega t + ce^{-Rt/L}.$$

Since $i(0) = i_0$ we obtain $c = i_0 + \frac{E_0 L \omega}{L^2 \omega^2 + R^2}$.

33. For $0 \le t \le 20$ the differential equation is $20\, di/dt + 2i = 120$. An integrating factor is $e^{t/10}$, so $(d/dt)[e^{t/10} i] = 6e^{t/10}$ and $i = 60 + c_1 e^{-t/10}$. If $i(0) = 0$ then $c_1 = -60$ and $i = 60 - 60e^{-t/10}$. For $t > 20$ the differential equation is $20\, di/dt + 2i = 0$ and $i = c_2 e^{-t/10}$. At $t = 20$ we want $c_2 e^{-2} = 60 - 60e^{-2}$ so that $c_2 = 60\left(e^2 - 1\right)$. Thus

$$\begin{cases} 60 - 60e^{-t/10}, & 0 \le t \le 20 \\ 60\left(e^2 - 1\right)e^{-t/10}, & t > 20. \end{cases}$$

Additional Linear Models

36. (a) Integrating $d^2 s/dt^2 = -g$ we get $v(t) = ds/dt = -gt + c$. From $v(0) = 300$ we find $c = 300$, and we are given $g = 32$, so the velocity is $v(t) = -32t + 300$.

(b) Integrating again and using $s(0) = 0$ we get $s(t) = -16t^2 + 300t$. The maximum height is attained when $v = 0$, that is, at $t_a = 9.375$. The maximum height will be $s(9.375) = 1406.25$ ft.

39. (a) The differential equation is first-order and linear. Letting $b = k/\rho$, the integrating factor is $e^{\int 3b\, dt/(bt+r_0)} = (r_0 + bt)^3$. Then

$$\frac{d}{dt}[(r_0 + bt)^3 v] = g(r_0 + bt)^3 \quad \text{and} \quad (r_0 + bt)^3 v = \frac{g}{4b}(r_0 + bt)^4 + c.$$

The solution of the differential equation is $v(t) = (g/4b)(r_0 + bt) + c(r_0 + bt)^{-3}$. Using $v(0) = 0$ we find $c = -gr_0^4/4b$, so that

$$v(t) = \frac{g}{4b}(r_0 + bt) - \frac{gr_0^4}{4b(r_0 + bt)^3} = \frac{g\rho}{4k}\left(r_0 + \frac{k}{\rho}t\right) - \frac{g\rho r_0^4}{4k(r_0 + kt/\rho)^3}.$$

(b) Integrating $dr/dt = k/\rho$ we get $r = kt/\rho + c$. Using $r(0) = r_0$ we have $c = r_0$, so $r(t) = kt/\rho + r_0$.

(c) If $r = 0.007\,\text{ft}$ when $t = 10\,\text{s}$, then solving $r(10) = 0.007$ for k/ρ, we obtain $k/\rho = -0.0003$ and $r(t) = 0.01 - 0.0003t$. Solving $r(t) = 0$ we get $t = 33.3$, so the raindrop will have evaporated completely at 33.3 seconds.

42. (a) The solution of the differential equation is $P(t) = c_1 e^{kt} + h/k$. If we let the initial population of fish be P_0 then $P(0) = P_0$ which implies that

$$c_1 = P_0 - \frac{h}{k} \quad \text{and} \quad P(t) = \left(P_0 - \frac{h}{k}\right) e^{kt} + \frac{h}{k}.$$

(b) For $P_0 > h/k$ all terms in the solution are positive. In this case $P(t)$ increases as time t increases. That is, $P(t) \to \infty$ as $t \to \infty$. For $P_0 = h/k$ the population remains constant for all time t:

$$P(t) = \left(\frac{h}{k} - \frac{h}{k}\right) e^{kt} + \frac{h}{k} = \frac{h}{k}.$$

For $0 < P_0 < h/k$ the coefficient of the exponential function is negative and so the function decreases as time t increases.

(c) Since the function decreases and is concave down, the graph of $P(t)$ crosses the t-axis. That is, there exists a time $T > 0$ such that $P(T) = 0$. Solving

$$\left(P_0 - \frac{h}{k}\right) e^{kT} + \frac{h}{k} = 0$$

for T shows that the time of extinction is

$$T = \frac{1}{k} \ln\left(\frac{h}{h - kP_0}\right).$$

45. (a) For $0 \le t < 4$, $6 \le t < 10$ and $12 \le t < 16$, no voltage is applied to the heart and $E(t) = 0$. At the other times, the differential equation is $dE/dt = -E/RC$. Separating variables, integrating, and solving for e, we get $E = ke^{-t/RC}$, subject to $E(4) = E(10) = E(16) = 12$. These intitial conditions yield, respectively, $k = 12e^{4/RC}$, $k = 12e^{10/RC}$, $k = 12e^{16/RC}$, and $k = 12e^{22/RC}$. Thus

$$
\begin{cases}
0, & 0 \le t < 4,\ 6 \le t < 10,\ 12 \le t < 16 \\
12e^{(4-t)/RC}, & 4 \le t < 6 \\
12e^{(10-t)/RC}, & 10 \le t < 12 \\
12e^{(16-t)/RC}, & 16 \le t < 18 \\
12e^{(22-t)/RC}, & 22 \le t < 24.
\end{cases}
$$

(b)

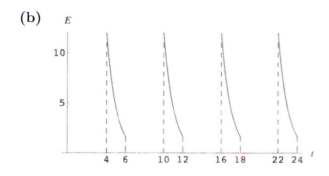

48. **(a)** We saw in part **(b)** of Problem 36 that the ascent time is $t_a = 9.375$. To find when the cannonball hits the ground we solve $s(t) = -16t^2 + 300t = 0$, getting a total time in flight of $t = 18.75\,\text{s}$. Thus, the time of descent is $t_d = 18.75 - 9.375 = 9.375$. The impact velocity is $v_i = v(18.75) = -300$, which has the same magnitude as the initial velocity.

(b) We saw in Problem 37 that the ascent time in the case of air resistance is $t_a = 9.162$. Solving $s(t) = 1,340,000 - 6,400t - 1,340,000e^{-0.005t} = 0$ we see that the total time of flight is $18.466\,\text{s}$. Thus, the descent time is $t_d = 18.466 - 9.162 = 9.304$. The impact velocity is $v_i = v(18.466) = -290.91$, compared to an initial velocity of $v_0 = 300$.

3.2 Nonlinear Models

The terminology and concepts listed below provide an outline of the main ideas encountered in this section. These can be useful when preparing for a quiz or test.

Terminology and Concepts

- logistic DE
- logistic growth
- harvesting
- Gompertz DE
- second-order chemical reaction

The basic skills listed below summarize the more mechanical types of problems encountered in the exercise set for this section.

Basic Skills

- set up and solve DEs involving the logistic model
- set up and solve DEs involving chemical reactions
- set up and solve DEs involving falling bodies subject to air resistance that is proportional to the square of the instantaneous velocity

Use of Computers

A scatter plot in two dimensions is a set of points in the plane. Both *Mathematica* and *Maple* have commands for plotting a set of points. In the examples below we plot the points $(-1, -2.7)$, $(-0.5, -2.5)$, $(0, -1.8)$, $(0.5, -1.6)$, $(1, -2.3)$, $(1.5, 0)$, $(2, 1.3)$, $(2.5, 1.9)$, $(3, 3.1)$.

$$\text{data} = \{\{\text{-1, -2.7}\}, \{\text{-0.5, -2.5}\}, \{0, \text{-1.8}\}, \{0.5, \text{-1.6}\},$$
$$\{1, \text{-2.3}\}, \{1.5, 0\}, \{2, 1.3\}, \{2.5, 1.9\}, \{3, 3.1\}\}$$
$$\text{grdata} = \text{ListPlot[data]}$$

(Mathematica)

The most commonly used options for **ListPlot** cause the dots to be enlarged or the dots to be connected with straight line segments. Although these two commands cannot be combined in a single call to **PlotStyle**, the same effect can be achieved as shown below.

$$\text{gr0} = \text{ListPlot[data, PlotStyle->PointSize[0.02]]}$$
$$\text{gr1} = \text{ListPlot[data, PlotJoined->True]}$$
$$\text{Show[gr0, gr1]}$$

(Mathematica)

In **Maple** a scatter plot is obtained using the **plots** package.

```
data:=[[-1,-2.7],[-0.5,-2.5],[0,-1.8],
    [0.5,-1.6],[1,-2.3],[1.5,0],[2,1.3],
    [2.5,1.9],[3,3.1]];
with(plots);
pointplot(data);
```

(Maple)

In **Maple** it is possible to show the points in a variety of symbols and symbol sizes. You can also connect the points with straight line segments. You cannot, however, do both in a single plot. The routine below shows how to accomplish this. It is assumed that the input data is as above and that the **plots** package has been activated.

```
gr0:=pointplot(data,style=point,          (Maple)
    symbol=circle);
gr1:=pointplot(data,style=line);
display([gr0,gr1]);
```

Algorithms that fit relatively simple functions to a set of data points are not difficult to develop but they tend to involve a large number of routine computations. This means that a CAS is particularly well suited to implementing a curve-fitting algorithm. The following routines use a least squares method to fit a line to the data set given above over the interval $[-1, 3]$. The resulting line is then plotted over the data points.

```
linfit = fit[data, {1, x}, x]              (Mathematica)
gr2 = Plot[linfit, {x, -1, 3}]
Show[gr0, gr2]
```

The format of **data** in the routines above is as a set of points with two coordinates. In **Maple**, to apply a least squares fit of a line to the set of points, we need to reformat the data as a set of x-coordinates, which we will call **xdata**, and a set of y-coordinates, which we will call **ydata**. To accomplish this we first use the **nops** command to count the number of data points. Then we activate the **stats** pacakage which contains commands for fitting a line to a set of points in the plane.

```
n:=nops(data);                             (Maple)
xdata:=[seq(data[i,1],i=1..n)];
ydata:=[seq(data[i,2],i=1..n)];
with(stats);
linfit:=fit[leastsquare[[x,y],y=a*x+b,{a,b}]]([xdata,ydata]);
gr3:=implicitplot(linfit,x=-1..3,y=-5..5);
display([gr0,gr3]);
```

In the above *Maple* routine, `implicitplot`, which is part of the **plots** package, is used because the result of the **fit** command is an equation rather than an expression or function. The range of the equation, in this case `y=-5..5`, is specified to be large enough to include all of the data points.

Logistic Equation

3. From $dP/dt = P\left(10^{-1} - 10^{-7}P\right)$ and $P(0) = 5000$ we obtain $P = 500/(0.0005 + 0.0995e^{-0.1t})$ so that $P \to 1{,}000{,}000$ as $t \to \infty$. If $P(t) = 500{,}000$ then $t = 52.9 \text{ months}$.

Modifications of the Logistic Model

6. Solving $P(5 - P) - \frac{25}{4} = 0$ for P we obtain the equilibrium solution $P = \frac{5}{2}$. For $P \neq \frac{5}{2}$, $dP/dt < 0$. Thus, if $P_0 < \frac{5}{2}$, the population becomes extinct (otherwise there would be another equilibrium solution.) Using separation of variables to solve the initial-value problem, we get

$$P(t) = [4P_0 + (10P_0 - 25)t]/[4 + (4P_0 - 10)t].$$

To find when the population becomes extinct for $P_0 < \frac{5}{2}$ we solve $P(t) = 0$ for t. We see that the time of extinction is $t = 4P_0/5(5 - 2P_0)$.

Chemical Reactions

9. Let $X = X(t)$ be the amount of C at time t and $dX/dt = k(120 - 2X)(150 - X)$. If $X(0) = 0$ and $X(5) = 10$, then

$$X(t) = \frac{150 - 150e^{180kt}}{1 - 2.5e^{180kt}},$$

where $k = .0001259$ and $X(20) = 29.3 \text{ grams}$. Now by L'Hôpital's rule, $X \to 60$ as $t \to \infty$, so that the amount of $A \to 0$ and the amount of $B \to 30$ as $t \to \infty$.

Additional nonlinear Models

12. To obtain the solution of this differential equation we use $h(t)$ from Problem 13 in Exercises 1.3. Then $h(t) = (A_w\sqrt{H} - 4cA_ht)^2/A_w^2$. Solving $h(t) = 0$ with $c = 0.6$ and the values from Problem 11 we see that the tank empties in 3035.79 seconds or 50.6 minutes.

15. (a) After separating variables we obtain

$$\frac{m\,dv}{mg - kv^2} = dt$$

$$\frac{1}{g}\frac{dv}{1 - (\sqrt{k}\,v/\sqrt{mg})^2} = dt$$

$$\frac{\sqrt{mg}}{\sqrt{k}\,g}\frac{\sqrt{k/mg}\,dv}{1 - (\sqrt{k}\,v/\sqrt{mg})^2} = dt$$

$$\sqrt{\frac{m}{kg}}\tanh^{-1}\frac{\sqrt{k}\,v}{\sqrt{mg}} = t + c$$

$$\tanh^{-1}\frac{\sqrt{k}\,v}{\sqrt{mg}} = \sqrt{\frac{kg}{m}}\,t + c_1.$$

Thus the velocity at time t is

$$v(t) = \sqrt{\frac{mg}{k}}\tanh\left(\sqrt{\frac{kg}{m}}t + c_1\right).$$

Setting $t = 0$ and $v = v_0$ we find $c_1 = \tanh^{-1}(\sqrt{k}\,v_0/\sqrt{mg})$.

(b) Since $\tanh t \to 1$ as $t \to \infty$, we have $v \to \sqrt{mg/k}$ as $t \to \infty$.

(c) Integrating the expression for $v(t)$ in part **(a)** we obtain an integral of the form $\int du/u$:

$$s(t) = \sqrt{\frac{mg}{k}}\int \tanh\left(\sqrt{\frac{kg}{m}}t + c_1\right)dt = \frac{m}{k}\ln\left[\cosh\left(\sqrt{\frac{kg}{m}}t + c_1\right)\right] + c_2.$$

Setting $t = 0$ and $s = 0$ we find $c_2 = -(m/k)\ln(\cosh c_1)$, where c_1 is given in part **(a)**.

18. (a) Writing the equation in the form $(x - \sqrt{x^2 + y^2})dx + y\,dy = 0$ we identify $M = x - \sqrt{x^2 + y^2}$ and $N = y$. Since M and N are both homogeneous functions of degree 1 we use the substitution $y = ux$. It follows that

$$\left(x - \sqrt{x^2 + u^2x^2}\right)dx + ux(u\,dx + x\,du) = 0$$

$$x\left[1 - \sqrt{1 + u^2} + u^2\right]dx + x^2u\,du = 0$$

$$-\frac{u\,du}{1 + u^2 - \sqrt{1 + u^2}} = \frac{dx}{x}$$

$$\frac{u\,du}{\sqrt{1 + u^2}\,(1 - \sqrt{1 + u^2})} = \frac{dx}{x}.$$

Letting $w = 1 - \sqrt{1 + u^2}$ we have $dw = -u\,du/\sqrt{1 + u^2}$ so that

$$-\ln\left|1 - \sqrt{1 + u^2}\right| = \ln|x| + c$$

$$\frac{1}{1 - \sqrt{1 + u^2}} = c_1 x$$

$$1 - \sqrt{1 + u^2} = -\frac{c_2}{x} \qquad (-c_2 = 1/c_1)$$

$$1 + \frac{c_2}{x} = \sqrt{1 + \frac{y^2}{x^2}}$$

$$1 + \frac{2c_2}{x} + \frac{c_2^2}{x^2} = 1 + \frac{y^2}{x^2}.$$

Solving for y^2 we have

$$y^2 = 2c_2 x + c_2^2 = 4\left(\frac{c_2}{2}\right)\left(x + \frac{c_2}{2}\right)$$

which is a family of parabolas symmetric with respect to the x-axis with vertex at $(-c_2/2, 0)$ and focus at the origin.

(b) Let $u = x^2 + y^2$ so that

$$\frac{du}{dx} = 2x + 2y\frac{dy}{dx}.$$

Then

$$y\frac{dy}{dx} = \frac{1}{2}\frac{du}{dx} - x$$

and the differential equation can be written in the form

$$\frac{1}{2}\frac{du}{dx} - x = -x + \sqrt{u} \quad \text{or} \quad \frac{1}{2}\frac{du}{dx} = \sqrt{u}.$$

Separating variables and integrating gives

$$\frac{du}{2\sqrt{u}} = dx$$

$$\sqrt{u} = x + c$$

$$u = x^2 + 2cx + c^2$$

$$x^2 + y^2 = x^2 + 2cx + c^2$$

$$y^2 = 2cx + c^2.$$

21. (a) With $c = 0.01$ the differential equation is $dP/dt = kP^{1.01}$. Separating variables and integrating we obtain

$$P^{-1.01} dP = k\, dt$$

$$\frac{P^{-0.01}}{-0.01} = kt + c_1$$

$$P^{-0.01} = -0.01kt + c_2$$

$$P(t) = (-0.01kt + c_2)^{-100}$$

$$P(0) = c_2^{-100} = 10$$

$$c_2 = 10^{-0.01}.$$

Then

$$P(t) = \frac{1}{(-0.01kt + 10^{-0.01})^{100}}$$

and, since P doubles in 5 months from 10 to 20,

$$P(5) = \frac{1}{(-0.01k(5) + 10^{-0.01})} = 20,$$

so

$$\left(-0.01k(5) + 10^{-0.01}\right)^{100} = \frac{1}{20}$$

$$-0.01k = \frac{\left[\left(\frac{1}{20}\right)^{1/100} - \left(\frac{1}{10}\right)^{1/100}\right]}{5}$$

$$= -0,001350.$$

Thus $P(t) = 1/\left(-0.001350t + 10^{-0.01}\right)^{100}$.

(b) Define $T = \left(\frac{1}{10}\right)^{1/100}/0.001350 \approx 724$ months $= 60$ years. As $t \to 724$ (from the left), $P \to \infty$.

(c) $P(50) = 1/\left[-0.001350(50) + 10^{-0.01}\right]^{100} \approx 12,839$ and

$$P(100) = 1/\left[-0.001350(100) + 10^{-0.01}\right]^{100} \approx 28,630,966$$

3.3 | Modeling with Systems of First-Order Differential Equations

The terminology and concepts listed below provide an outline of the main ideas encountered in this section. These can be useful when preparing for a quiz or test.

Terminology and Concepts

- systems of first-order DEs – linear and nonlinear

- radioactive decay series

- mixture models involving a system of DEs

- Lotka-Volterra predator-prey model

- competition models

- electrical networks involving a system of first-order DEs

The basic skills listed below summarize the more mechanical types of problems encountered in the exercise set for this section.

Basic Skills

- solve a system of DEs where one of the equations involves only a single unknown

- construct a mathematical model for mixtures involving multiple tanks

- use a computer program to numerically analyze solutions of a system of DEs

Use of Computers

There are a number of computer programs that can solve a DE or system of DEs numerically and graph the resulting numerical solution curves. We show below how *Mathematica* and *Maple* can produce graphs of numerical solutions.

In *Mathematica* **NDSolve** is used to obtain a numerical solution of an IVP which can then be graphed.

```
Clear[y]                                                    (Mathematica)
sol=Flatten[NDSolve[{y'[x]==Exp[x/2]Sin[x^2], y[0]==0}, y, {x, -5, 5}]];
y[x_]:=Evaluate[y[x]/.sol]
y[x]
Plot[y[x], {x, -5, 5}]
```

Note that the interval $[-5, 5]$ specified in the **Plot** command can be no larger than the interval used in the **NDSolve** command. Following is an example of how *Mathematica* can be used to obtain a numerical solution of a system of DEs with initial conditions.

```
Clear[y]
sol=Flatten[NDSolve[{x'[t]== -0.3x[t] + 0.1x[t]y[t],
y'[t]==0.2y[t] - 0.1x[t]y[t], x[0]==6, y[0]==4}, {x, y}, {t, -2, 2}]];
x[t_]:=Evaluate[x[t]/.sol]
y[t_]:=Evaluate[y[t]/.sol]
x[t]
y[t]
Plot[{x[t], y[t]}, {t, -2, 2}]
```

In *Maple* there are two ways to plot the numerical solution curve of an IVP. The first way shown below returns a procedure that can be used with the **plots** package to graph the solution. It is also possible to define a function that can evaluate the solution at any point.

```
sol:=dsolve({diff(y(x),x)=exp(x/t)*sin(x^2),     (Maple)
y(0)=0},{y(x)}, type=numeric);
with(plots);
odeplots(sol,[x,y(x)],-5..5);
y1:=x1->eval(y(x),sol(x1));
y1(2.67);
```

The second way to plot a numerical solution of the DE uses the **DEtools** package.

```
with(DEtools);                                   (Maple)
DEplot(diff(y(x),x)=exp(x/2)*sin(x^2),y(x),-5..5,
    [[y(0)=0]],arrows=NONE,stepsize=0.05);
```

The **arrows** option suppresses a direction field in the background of the graph and the **stepsize** option smooths the graph out.

Maple can also graph the numerical solutions of a system of DEs with initial conditions. In this case we first specify the DEs and initial conditions before calling the **DEtools** package.

```
eq1:=diff(x(t),t)=-0.3*x(t)+0.1*x(t)*y(t);       (Maple)
eq2:=diff(y(t),t)=0.2*y(t)-0.1*x(t)*y(t);
with(DEtools);
gr1:=DEplot({eq1,eq2},[x(t),y(t)],-5..5,[[x(0)=6,y(0)=4]],
    scene=[t,x]):
gr2:=DEplot({eq1,eq2},[x(t),y(t)],-5..5,[[x(0)=6,y(0)=4]],
    scene=[t,y]):
display([gr1,gr2]);
```

Note the use of colons, rather than the usual **semicolons**, in the lines above defining **gr1** and **gr2**. This prevents the display of a very long list of uninteresting expressions.

Radioactive Series

3. The amounts x and y are the same at about $t = 5$ days. The amounts x and z are the same at about $t = 20$ days. The amounts y and z are the same at about $t = 147$ days. The time when y and z are the same makes sense because most of A and half of B are gone, so half of C should have been formed.

Mixtures

6. Let x_1, x_2, and x_3 be the amounts of salt in tanks A, B, and C, respectively, so that

$$x_1' = \frac{1}{100}x_2 \cdot 2 - \frac{1}{100}x_1 \cdot 6 = \frac{1}{50}x_2 - \frac{3}{50}x_1$$

$$x_2' = \frac{1}{100}x_1 \cdot 6 + \frac{1}{100}x_3 - \frac{1}{100}x_2 \cdot 2 - \frac{1}{100}x_2 \cdot 5 = \frac{3}{50}x_1 - \frac{7}{100}x_2 + \frac{1}{100}x_3$$

$$x_3' = \frac{1}{100}x_2 \cdot 5 - \frac{1}{100}x_3 - \frac{1}{100}x_3 \cdot 4 = \frac{1}{20}x_2 - \frac{1}{20}x_3.$$

Predator-Prey Models

9. Zooming in on the graph it can be seen that the populations are first equal at about $t = 5.6$. The approximate periods of x and y are both 45.

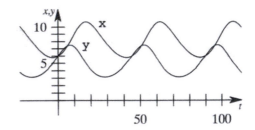

Networks

12. By Kirchhoff's first law we have $i_1 = i_2 + i_3$. By Kirchhoff's second law, on each loop we have $E(t) = Li_1' + R_1 i_2$ and $E(t) = Li_1' + R_2 i_3 + q/C$ so that $q = CR_1 i_2 - CR_2 i_3$. Then $i_3 = q' = CR_1 i_2' - CR_2 i_3$ so that the system is

$$Li_2' + Li_3' + R_1 i_2 = E(t)$$

$$-R_1 i_2' + R_2 i_3' + \frac{1}{C}i_3 = 0.$$

Additional Nonlinear Models

15. We first note that $s(t) + i(t) + r(t) = n$. Now the rate of change of the number of susceptible persons, $s(t)$, is proportional to the number of contacts between the number of people infected and the number who are susceptible; that is, $ds/dt = -k_1 si$. We use $-k_1 < 0$ because $s(t)$ is decreasing. Next, the rate of change of the number of persons who have recovered is proportional to the number infected; that is, $dr/dt = k_2 i$ where $k_2 > 0$ since r is increasing. Finally, to obtain di/dt we use

$$\frac{d}{dt}(s + i + r) = \frac{d}{dt}n = 0.$$

This gives

$$\frac{di}{dt} = -\frac{dr}{dt} - \frac{ds}{dt} = -k_2 i + k_1 si.$$

The system of differential equations is then

$$\frac{ds}{dt} = -k_1 si$$

$$\frac{di}{dt} = -k_2 i + k_1 si$$

$$\frac{dr}{dt} = k_2 i.$$

A reasonable set of initial conditions is $i(0) = i_0$, the number of infected people at time 0, $s(0) = n - i_0$, and $r(0) = 0$.

3.R   Chapter 3 in Review

3. From $\dfrac{dP}{dt} = 0.018P$ and $P(0) = 4$ billion we obtain $P = 4e^{0.018t}$ so that $P(45) = 8.99$ billion.

6. From $V\,dC/dt = kA(C_s - C)$ and $C(0) = C_0$ we obtain $C = C_s + (C_0 - C_s)e^{-kAt/V}$.

9. We first solve $(1 - t/10)di/dt + 0.2i = 4$. Separating variables we obtain $di/(40 - 2i) = dt/(10 - t)$. Then

$$-\frac{1}{2}\ln|40 - 2i| = -\ln|10 - t| + c \quad \text{or} \quad \sqrt{40 - 2i} = c_1(10 - t).$$

Since $i(0) = 0$ we must have $c_1 = 2/\sqrt{10}$. Solving for i we get $i(t) = 4t - \frac{1}{5}t^2$, $0 \le t < 10$. For $t \ge 10$ the equation for the current becomes

$$i(t) = \begin{cases} 4t - \frac{1}{5}t^2, & 0 \le t < 10 \\ 20, & t \ge 10. \end{cases}$$

The graph of $i(t)$ is given in the figure.

12. In tank A the salt input is

$$\left(7\,\frac{\text{gal}}{\text{min}}\right)\left(2\,\frac{\text{lb}}{\text{gal}}\right)+\left(1\,\frac{\text{gal}}{\text{min}}\right)\left(\frac{x_2}{100}\,\frac{\text{lb}}{\text{gal}}\right)=\left(14+\frac{1}{100}x_2\right)\frac{\text{lb}}{\text{min}}.$$

The salt output is

$$\left(3\,\frac{\text{gal}}{\text{min}}\right)\left(\frac{x_1}{100}\,\frac{\text{lb}}{\text{gal}}\right)+\left(5\,\frac{\text{gal}}{\text{min}}\right)\left(\frac{x_1}{100}\,\frac{\text{lb}}{\text{gal}}\right)=\frac{2}{25}x_1\,\frac{\text{lb}}{\text{min}}.$$

In tank B the salt input is

$$\left(5\,\frac{\text{gal}}{\text{min}}\right)\left(\frac{x_1}{100}\,\frac{\text{lb}}{\text{gal}}\right)=\frac{1}{20}x_1\,\frac{\text{lb}}{\text{min}}.$$

The salt output is

$$\left(1\,\frac{\text{gal}}{\text{min}}\right)\left(\frac{x_2}{100}\,\frac{\text{lb}}{\text{gal}}\right)+\left(4\,\frac{\text{gal}}{\text{min}}\right)\left(\frac{x_2}{100}\,\frac{\text{lb}}{\text{gal}}\right)=\frac{1}{20}x_2\,\frac{\text{lb}}{\text{min}}.$$

The system of differential equations is then

$$\frac{dx_1}{dt}=14+\frac{1}{100}x_2-\frac{2}{25}x_1$$

$$\frac{dx_2}{dt}=\frac{1}{20}x_1-\frac{1}{20}x_2.$$

15. (a) From the third differential equation in the statement of the problem in the text we have

$$\frac{dK}{dt}=-(\lambda_1+\lambda_2)K \qquad \text{so} \qquad K(t)=c_1e^{-(\lambda_1+\lambda_2)t}.$$

Since $K(0)=K_0$ we have $K(t)=K_0e^{-(\lambda_1+\lambda_2)t}$. Next,

$$\frac{dC}{dt}=\lambda_1 K=\lambda_1 K_0 e^{-(\lambda_1+\lambda_2)t} \qquad \text{so} \qquad C(t)=-\frac{\lambda_1}{\lambda_1+\lambda_2}+c_2.$$

Since $C(0)=0$ we have

$$c_2=\frac{\lambda_1}{\lambda_1+\lambda_2}K_0 \qquad \text{and} \qquad C(t)=\frac{\lambda_1}{\lambda_1+\lambda_2}K_0\left[1-e^{-(\lambda_1+\lambda_2)t}\right].$$

Finally

$$\frac{dA}{dt}=\lambda_2 K=\lambda_2 K_0 e^{-(\lambda_1+\lambda_2)t} \qquad \text{so} \qquad A(t)=-\frac{\lambda_1}{\lambda_1+\lambda_2}K_0 e^{-(\lambda_1+\lambda_2)t}+c_3.$$

Since $A(0)=0$

$$c_3=\frac{\lambda_2}{\lambda_1+\lambda_2}K_0 \qquad \text{and} \qquad A(t)=\frac{\lambda_2}{\lambda_1+\lambda_2}K_0\left[1-e^{-(\lambda_1+\lambda_2)t}\right].$$

(b) Since $\lambda_1 + \lambda_2 = 5.34 \times 10^{-10}$ we have

$$K(t) = K_0 e^{-0.000\,000\,000\,534t} = \frac{1}{2}K_0 \quad \text{and} \quad t = \frac{\ln\frac{1}{2}}{-0.000\,000\,000\,534} \approx 1.3 \times 10^9 \text{ years.}$$

(c) Since

$$\lim_{t\to\infty} C(t) = \frac{\lambda_1}{\lambda_1 + \lambda_2} \lim_{t\to\infty} K_0\left[1 - e^{-(\lambda_1+\lambda_2)t}\right] = \frac{\lambda_1}{\lambda_1 + \lambda_2}K_0$$

and

$$\lim_{t\to\infty} A(t) = \frac{\lambda_1}{\lambda_1 + \lambda_2} \lim_{t\to\infty} K_0\left[1 - e^{-(\lambda_2+\lambda_2)t}\right] = \frac{\lambda_2}{\lambda_1 + \lambda_2}K_0,$$

we have

$$\lim_{t\to\infty} C(t) = \frac{4.7526 \times 10^{-10}}{5.34 \times 10^{-10}}K_0 = 0.89K_0 \quad \text{or} \quad 89\%$$

and

$$\lim_{t\to\infty} A(t) = \frac{0.5874 \times 10^{-10}}{5.34 \times 10^{-10}}K_0 = 0.11K_0 \quad \text{or} \quad 11\%.$$

4 Higher-Order Differential Equations

4.1 | Preliminary Theory – Linear Equations

The terminology and concepts listed below provide an outline of the main ideas encountered in this section. These can be useful when preparing for a quiz or test.

Terminology and Concepts

- linear nth-order initial-value problem
- existence-uniqueness theorem for linear nth-order IVPs
- boundary-value problems for linear nth-order DEs
- no existence-uniqueness theorem for linear nth-order boundary-value problems
- homogeneous linear nth-order DEs
- representation of a DE using differential operators
- superposition principles
- linear dependence and linear independence of a set of functions
- Wronskian
- fundamental set of solutions
- general solution of a linear nth-order DE
- particular solution of a nonhomogeneous linear nth-order DE
- complementary function for a nonhomogeneous linear nth-order DE

The basic skills listed below summarize the more mechanical types of problems encountered in the exercise set for this section.

Basic Skills

- given the general solution of a linear nth-order DE, find a solution satisfying n initial conditions
- given a two-parameter family of solutions of a linear second-order DE, determine a member or members of the family satisfying specified boundary conditions (or show that no solution exists)
- determine whether a given set of functions is linearly independent on an interval
- show that a given set of functions forms a fundamental set of solutions on an interval

74

4.1.1 Initial-Value and Boundary-Value Problems

3. From $y = c_1 x + c_2 x \ln x$ we find $y' = c_1 + c_2(1 + \ln x)$. Then $y(1) = c_1 = 3$, $y'(1) = c_1 + c_2 = -1$ so that $c_1 = 3$ and $c_2 = -4$. The solution is $y = 3x - 4x \ln x$.

6. In this case we have $y(0) = c_1 = 0$, $y'(0) = 2c_2 \cdot 0 = 0$ so $c_1 = 0$ and c_2 is arbitrary. Two solutions are $y = x^2$ and $y = 2x^2$.

9. Since $a_2(x) = x - 2$ and $x_0 = 0$ the problem has a unique solution for $-\infty < x < 2$.

12. In this case we have $y(0) = c_1 = 1$, $y'(1) = 2c_2 = 6$ so that $c_1 = 1$ and $c_2 = 3$. The solution is $y = 1 + 3x^2$.

4.1.2 Homogeneous Equations

15. Since $(-4)x + (3)x^2 + (1)(4x - 3x^2) = 0$ the set of functions is linearly dependent.

18. Since $(1)\cos 2x + (1)1 + (-2)\cos^2 x = 0$ the set of functions is linearly dependent.

21. Suppose $c_1(1 + x) + c_2 x + c_3 x^2 = 0$. Then $c_1 + (c_1 + c_2)x + c_3 x^2 = 0$ and so $c_1 = 0$, $c_1 + c_2 = 0$, and $c_3 = 0$. Since $c_1 = 0$ we also have $c_2 = 0$. Thus, the set of functions is linearly independent.

24. The functions satisfy the differential equation and are linearly independent since

$$W(\cosh 2x, \sinh 2x) = 2$$

for $-\infty < x < \infty$. The general solution is

$$y = c_1 \cosh 2x + c_2 \sinh 2x.$$

27. The functions satisfy the differential equation and are linearly independent since

$$W\left(x^3, x^4\right) = x^6 \neq 0$$

for $0 < x < \infty$. The general solution on this interval is

$$y = c_1 x^3 + c_2 x^4.$$

30. The functions satisfy the differential equation and are linearly independent since

$$W(1, x, \cos x, \sin x) = 1$$

for $-\infty < x < \infty$. The general solution on this interval is

$$y = c_1 + c_2 x + c_3 \cos x + c_4 \sin x.$$

4.1.3 Nonhomogeneous Equations

33. The functions $y_1 = e^{2x}$ and $y_2 = xe^{2x}$ form a fundamental set of solutions of the associated homogeneous equation, and $y_p = x^2 e^{2x} + x - 2$ is a particular solution of the nonhomogeneous equation.

36. **(a)** $y_{p_1} = 5$ **(b)** $y_{p_2} = -2x$ **(c)** $y_p = y_{p_1} + y_{p_2} = 5 - 2x$ **(d)** $y_p = \frac{1}{2} y_{p_1} - 2y_{p_2} = \frac{5}{2} + 4x$

| **4.2** | Reduction of Order |

The terminology and concepts listed below provide an outline of the main ideas encountered in this section. These can be useful when preparing for a quiz or test.

Terminology and Concepts

- reduction of order

The basic skills listed below summarize the more mechanical types of problems encountered in the exercise set for this section.

Basic Skills

- given a solution y_1 of a linear second-order homogeneous DE, find a second solution y_2 using reduction of order

- given a solution y_1 of a linear second-order homogeneous DE, find a second solution y_2 using formula (5) in the text

In Problems 3 and 6 we use reduction of order to find a second solution. In Problems 9-18 we use formula (5) from the text.

3. Define $y = u(x) \cos 4x$ so

$$y' = -4u \sin 4x + u' \cos 4x, \quad y'' = u'' \cos 4x - 8u' \sin 4x - 16u \cos 4x$$

and

$$y'' + 16y = (\cos 4x)u'' - 8(\sin 4x)u' = 0 \quad \text{or} \quad u'' - 8(\tan 4x)u' = 0.$$

If $w = u'$ we obtain the linear first-order equation $w' - 8(\tan 4x)w = 0$ which has the integrating factor $e^{-8\int \tan 4x\,dx} = \cos^2 4x$. Now

$$\frac{d}{dx}[(\cos^2 4x)w] = 0 \quad \text{gives} \quad (\cos^2 4x)w = c.$$

Therefore $w = u' = c\sec^2 4x$ and $u = c_1 \tan 4x$. A second solution is $y_2 = \tan 4x \cos 4x = \sin 4x$.

6. Define $y = u(x)e^{5x}$ so

$$y' = 5e^{5x}u + e^{5x}u', \quad y'' = e^{5x}u'' + 10e^{5x}u' + 25e^{5x}u$$

and

$$y'' - 25y = e^{5x}(u'' + 10u') = 0 \quad \text{or} \quad u'' + 10u' = 0.$$

If $w = u'$ we obtain the linear first-order equation $w' + 10w = 0$ which has the integrating factor $e^{10\int dx} = e^{10x}$. Now

$$\frac{d}{dx}[e^{10x}w] = 0 \quad \text{gives} \quad e^{10x}w = c.$$

Therefore $w = u' = ce^{-10x}$ and $u = c_1 e^{-10x}$. A second solution is $y_2 = e^{-10x}e^{5x} = e^{-5x}$.

9. Identifying $P(x) = -7/x$ we have

$$y_2 = x^4 \int \frac{e^{-\int(-7/x)\,dx}}{x^8}\,dx = x^4 \int \frac{1}{x}\,dx = x^4 \ln|x|.$$

A second solution is $y_2 = x^4 \ln|x|$.

12. Identifying $P(x) = 0$ we have

$$y_2 = x^{1/2} \ln x \int \frac{e^{-\int 0\,dx}}{x(\ln x)^2}\,dx = x^{1/2} \ln x \left(-\frac{1}{\ln x}\right) = -x^{1/2}.$$

A second solution is $y_2 = x^{1/2}$.

15. Identifying $P(x) = 2(1 + x)/\left(1 - 2x - x^2\right)$ we have

$$y_2 = (x + 1) \int \frac{e^{-\int 2(1+x)\,dx/\left(1-2x-x^2\right)}}{(x + 1)^2}\,dx = (x + 1) \int \frac{e^{\ln\left(1-2x-x^2\right)}}{(x + 1)^2}\,dx$$

$$= (x + 1) \int \frac{1 - 2x - x^2}{(x + 1)^2}\,dx = (x + 1) \int \left[\frac{2}{(x + 1)^2} - 1\right]dx$$

$$= (x + 1)\left[-\frac{2}{x + 1} - x\right] = -2 - x^2 - x.$$

A second solution is $y_2 = x^2 + x + 2$.

18. Define $y = u(x) \cdot 1$ so

$$y' = u', \quad y'' = u'' \quad \text{and} \quad y'' + y' = u'' + u' = 1.$$

If $w = u'$ we obtain the linear first-order equation $w' + w = 1$ which has the integrating factor $e^{\int dx} = e^x$. Now

$$\frac{d}{dx}[e^x w] = e^x \quad \text{gives} \quad e^x w = e^x + c.$$

Therefore $w = u' = 1 + ce^{-x}$ and $u = x + c_1 e^{-x} + c_2$. The general solution is

$$y = u = x + c_1 e^{-x} + c_2.$$

4.3 │ Homogeneous Linear Equations with Constant Coefficients

The terminology and concepts listed below provide an outline of the main ideas encountered in this section. These can be useful when preparing for a quiz or test.

Terminology and Concepts

- auxiliary equation

The basic skills listed below summarize the more mechanical types of problems encountered in the exercise set for this section.

Basic Skills

- find the general solution of a homogeneous linear DE with constant coefficients when the roots of the auxiliary equation are real and distinct, real and repeated, conjugate complex, and repeated conjugate complex

- find the solution of an nth-order homogeneous linear DE with constant coefficients and n initial conditions

Factorization of Polynomials   In Section 4.3 in the text the solution method for homogeneous linear DEs with constant coefficents relies on your ability to solve polynomial equations. For example, you know that a quadratic polynomial equation

$$am^2 + bm + c = 0, \quad a \neq 0, \ a, b, c \text{ real constants},$$

can always be solved using the quadratic formula. The two roots m_1 and m_2 are

$$m_1 = \frac{-b + \sqrt{b^2 - 4ac}}{2a} \quad \text{and} \quad m_2 = \frac{-b - \sqrt{b^2 - 4ac}}{2a}.$$

The roots m_1 and m_2 are also called the **zeros** of the polynomial $am^2 + bm + c$. Note that m_1 and m_2 are real and distinct $(m_1 \neq m_2)$ if $b^2 - 4ac > 0$, real and equal $(m_1 = m_2)$ if $b^2 - 4ac = 0$, and finally, m_1 and m_2 are complex conjugates $(m_1 = \alpha + i\beta, m_1 = \alpha - i\beta, \alpha$ and β are positive real numbers) if $b^2 - 4ac < 0$. The Factor Theorem of Algebra states that finding zeros of a polynomial and factoring the polynomial are equivalent problems. Thus, in the case of a quadratic polynomial $am^2 + bm + c$ we can write

$$am^2 + bm + c = a(m - m_1)(m - m_2).$$

A polynomial of degree n, $a_n m^n + a_{n-1} m^{n-1} + \cdots + a_1 m + a_0$ with real coefficients has n zeros (counting multiplicities) and hence can be factored as

$$a_n m^n + a_{n-1} m^{n-1} + \cdots + a_1 m + a_0 = a_n(m - m_1)(m - m_2) \cdots (m - m_n).$$

Finding the zeros requires that we find roots of a polynomial equation of degree > 2, and *that* is the fundamental problem here. Analogous to quadratic equations, cubic and quartic polynomial equations can be solved by (complicated) formulas involving radicals, but there exist no such formulas for solving polynomial equations of degree $n > 4$. But if the polynomial has *integer* coefficients and *if p/q is a rational zero* (p and q are integers and $q \neq 0$), then it can be proved that

$$p \text{ must be an integer factor of } a_0,$$

and

$$q \text{ must be an integer factor of } a_n.$$

By forming all possible quotients of each factor of a_0 to each factor of a_n, we can construct a list of all *possible* rational zeros. For example, consider the fourth-degree polynomial equation $3m^4 - 10m^3 - 3m^2 + 8m - 2 = 0$. We identify $a_0 = -2$ and $a_n = 3$ and list all integer factors of a_0 and a_n, respectively:

$$p : \pm 1, \pm 2 \qquad \text{and} \qquad q : \pm 1, \pm 3.$$

The eight possible rational roots of the equation are then

$$\frac{p}{q} : -1, 1, -\frac{1}{3}, \frac{1}{3}, -2, 2, -\frac{2}{3}, \frac{2}{3}.$$

To determine which, if any, of these numbers are roots, we could use brute force substitution of a number p/q into the polynomial to see whether it satisfies the equation. We could also use long division. If the term $m - p/q$ divides the polynomial evenly, that is, the remainder r is zero, then $m - p/q$ is a factor and hence p/q is a root. A short-hand version of long division, called **synthetic division**, is an efficient way of proceeding. We begin by testing -1 by synthetically dividing by $m - (-1) = m + 1$:

Since the remainder upon division is $r = 0$, we can conclude $m + 1$ is a factor of the polynomial and so -1 is a root. Now the division shows that

$$3m^4 - 10m^3 - 3m^2 + 8m - 2 = (m + 1)(3m^3 - 13m^2 + 10m - 2).$$

Observe that if the original polynomial has additional rational zeros they must be zeros of the second, or cubic, factor $3m^3 - 13m^2 + 10m - 2$. Thus we continue the testing by dividing this factor by $m - 1$:

$$
\begin{array}{r|rrrr}
1 & 3 & -13 & 10 & -2 \\
  &   & 3 & -10 & 0 \\
\hline
  & 3 & -10 & 0 & \boxed{-2 = r}
\end{array}
$$

The foregoing division shows that $m - 1$ is not a factor since the remainder r is not zero, and so 1 is not a rational root. Checking $\frac{1}{3}$ we have

$$
\begin{array}{r|rrrr}
\frac{1}{3} & 3 & -13 & 10 & -2 \\
  &   & 1 & -4 & 2 \\
\hline
  & 3 & -12 & 6 & \boxed{0 = r}
\end{array}
$$

Thus, $\frac{1}{3}$ is a root, and we now have the factorization

$$
\begin{aligned}
3m^4 - 10m^3 - 3m^2 + 8m - 2 &= (m + 1)\left(m - \frac{1}{3}\right)(3m^2 - 12m + 6) \\
&= (m + 1)\left(m - \frac{1}{3}\right)(3)(m^2 - 4m + 2) \\
&= (m + 1)(3m - 1)(m^2 - 4m + 2).
\end{aligned}
$$

Since the factor $m^2 - 4m + 2$ discovered by the division is quadratic, the remaining roots, $2 + \sqrt{2}$ and $2 - \sqrt{2}$, can be determined from the quadratic formula. Therefore, the original polynomial equation has four real roots: two rational and two irrational.

3. From $m^2 - m - 6 = 0$ we obtain $m = 3$ and $m = -2$ so that $y = c_1 e^{3x} + c_2 e^{-2x}$.

6. From $m^2 - 10m + 25 = 0$ we obtain $m = 5$ and $m = 5$ so that $y = c_1 e^{5x} + c_2 x e^{5x}$.

9. From $m^2 + 9 = 0$ we obtain $m = 3i$ and $m = -3i$ so that $y = c_1 \cos 3x + c_2 \sin 3x$.

12. From $2m^2 + 2m + 1 = 0$ we obtain $m = -1/2 \pm i/2$ so that

$$y = e^{-x/2}[c_1 \cos(x/2) + c_2 \sin(x/2)].$$

15. From $m^3 - 4m^2 - 5m = 0$ we obtain $m = 0$, $m = 5$, and $m = -1$ so that

$$y = c_1 + c_2 e^{5x} + c_3 e^{-x}.$$

18. From $m^3 + 3m^2 - 4m - 12 = 0$ we obtain $m = -2$, $m = 2$, and $m = -3$ so that

$$y = c_1 e^{-2x} + c_2 e^{2x} + c_3 e^{-3x}.$$

21. From $m^3 + 3m^2 + 3m + 1 = 0$ we obtain $m = -1$, $m = -1$, and $m = -1$ so that

$$y = c_1 e^{-x} + c_2 x e^{-x} + c_3 x^2 e^{-x}.$$

24. From $m^4 - 2m^2 + 1 = 0$ we obtain $m = 1$, $m = 1$, $m = -1$, and $m = -1$ so that

$$y = c_1 e^x + c_2 x e^x + c_3 e^{-x} + c_4 x e^{-x}.$$

27. From $m^5 + 5m^4 - 2m^3 - 10m^2 + m + 5 = 0$ we obtain $m = -1$, $m = -1$, $m = 1$, and $m = 1$, and $m = -5$ so that

$$u = c_1 e^{-r} + c_2 r e^{-r} + c_3 e^r + c_4 r e^r + c_5 e^{-5r}.$$

30. From $m^2 + 1 = 0$ we obtain $m = \pm i$ so that $y = c_1 \cos \theta + c_2 \sin \theta$. If $y(\pi/3) = 0$ and $y'(\pi/3) = 2$ then

$$\frac{1}{2} c_1 + \frac{\sqrt{3}}{2} c_2 = 0$$

$$-\frac{\sqrt{3}}{2} c_1 + \frac{1}{2} c_2 = 2,$$

so $c_1 = -\sqrt{3}$, $c_2 = 1$, and $y = -\sqrt{3} \cos \theta + \sin \theta$.

33. From $m^2 + m + 2 = 0$ we obtain $m = -1/2 \pm \sqrt{7}\, i/2$ so that

$$y = e^{-x/2}[c_1 \cos(\sqrt{7}\, x/2) + c_2 \sin(\sqrt{7}\, x/2)].$$

If $y(0) = 0$ and $y'(0) = 0$ then $c_1 = 0$ and $c_2 = 0$ so that $y = 0$.

36. From $m^3 + 2m^2 - 5m - 6 = 0$ we obtain $m = -1$, $m = 2$, and $m = -3$ so that

$$y = c_1 e^{-x} + c_2 e^{2x} + c_3 e^{-3x}.$$

If $y(0) = 0$, $y'(0) = 0$, and $y''(0) = 1$ then

$$c_1 + c_2 + c_3 = 0, \quad -c_1 + 2c_2 - 3c_3 = 0, \quad c_1 + 4c_2 + 9c_3 = 1,$$

so $c_1 = -1/6$, $c_2 = 1/15$, $c_3 = 1/10$, and

$$y = -\frac{1}{6}e^{-x} + \frac{1}{15}e^{2x} + \frac{1}{10}e^{-3x}.$$

39. From $m^2 + 1 = 0$ we obtain $m = \pm i$ so that $y = c_1 \cos x + c_2 \sin x$ and $y' = -c_1 \sin x + c_2 \cos x$. From $y'(0) = c_1(0) + c_2(1) = c_2 = 0$ and $y'(\pi/2) = -c_1(1) = 0$ we find $c_1 = c_2 = 0$. A solution of the boundary-value problem is $y = 0$.

42. The auxiliary equation is $m^2 - 1 = 0$ which has roots -1 and 1. By (10) the general solution is $y = c_1 e^x + c_2 e^{-x}$. By (11) the general solution is $y = c_1 \cosh x + c_2 \sinh x$. For $y = c_1 e^x + c_2 e^{-x}$ the boundary conditions imply $c_1 + c_2 = 1$, $c_1 e - c_2 e^{-1} = 0$. Solving for c_1 and c_2 we find $c_1 = 1/(1 + e^2)$ and $c_2 = e^2/(1 + e^2)$ so $y = e^x/(1 + e^2) + e^2 e^{-x}/(1 + e^2)$. For $y = c_1 \cosh x + c_2 \sinh x$ the boundary conditions imply $c_1 = 1$, $c_2 = -\tanh 1$, so $y = \cosh x - (\tanh 1) \sinh x$.

45. The auxiliary equation should have a pair of complex roots $\alpha \pm \beta i$ where $\alpha < 0$, so that the solution has the form $e^{\alpha x}(c_1 \cos \beta x + c_2 \sin \beta x)$. Thus, the differential equation is **(e)**.

48. The differential equation should have the form $y'' + k^2 y = 0$ where $k = 2$ so that the period of the solution is π. Thus, the differential equation is **(b)**.

51. We have $m(m - 2) = m^2 - 2m$, so the differential equation is $y'' - 2y' = 0$.

54. We have $(m - 7)(m + 7) = m^2 - 49$, so the differential equation is $y'' - 49y = 0$.

57. We have $m^2(m - 8) = m3 - 8m^2$, so the differential equation is $y''' - 8y'' = 0$.

4.4 Undetermined Coefficients—Superposition Approach

The terminology and concepts listed below provide an outline of the main ideas encountered in this section. These can be useful when preparing for a quiz or test.

Terminology and Concepts

- undetermined coefficients

- superposition principle

- multiplication rule

The basic skills listed below summarize the more mechanical types of problems encountered in the exercise set for this section.

Basic Skills

- use undetermined coefficients with the superposition approach to find a particular solution of a nonhomogeneous linear DE with constant coefficients when (1) no function in the assumed particular solution is a solution of the associated homogeneous DE, and (2) a function in the assumed particular solution is also a solution of the associated homogeneous DE

- recognize when the input function (right-hand side of the DE) is appropriate for the method of undetermined coefficients.

3. From $m^2 - 10m + 25 = 0$ we find $m_1 = m_2 = 5$. Then $y_c = c_1 e^{5x} + c_2 x e^{5x}$ and we assume that $y_p = Ax + B$. Substituting into the differential equation we obtain $25A = 30$ and $-10A + 25B = 3$. Then $A = \frac{6}{5}$, $B = \frac{3}{5}$, $y_p = \frac{6}{5}x + \frac{3}{5}$, and

$$y = c_1 e^{5x} + c_2 x e^{5x} + \frac{6}{5}x + \frac{3}{5}.$$

6. From $m^2 - 8m + 20 = 0$ we find $m_1 = 4 + 2i$ and $m_2 = 4 - 2i$. Then $y_c = e^{4x}(c_1 \cos 2x + c_2 \sin 2x)$ and we assume $y_p = Ax^2 + Bx + C + (Dx + E)e^x$. Substituting into the differential equation we obtain

$$2A - 8B + 20C = 0$$
$$-6D + 13E = 0$$
$$-16A + 20B = 0$$
$$13D = -26$$
$$20A = 100.$$

Then $A = 5$, $B = 4$, $C = \frac{11}{10}$, $D = -2$, $E = -\frac{12}{13}$, $y_p = 5x^2 + 4x + \frac{11}{10} + \left(-2x - \frac{12}{13}\right)e^x$ and

$$y = e^{4x}(c_1 \cos 2x + c_2 \sin 2x) + 5x^2 + 4x + \frac{11}{10} + \left(-2x - \frac{12}{13}\right)e^x.$$

9. From $m^2 - m = 0$ we find $m_1 = 1$ and $m_2 = 0$. Then $y_c = c_1 e^x + c_2$ and we assume $y_p = Ax$. Substituting into the differential equation we obtain $-A = -3$. Then $A = 3$, $y_p = 3x$ and $y = c_1 e^x + c_2 + 3x$.

12. From $m^2 - 16 = 0$ we find $m_1 = 4$ and $m_2 = -4$. Then $y_c = c_1 e^{4x} + c_2 e^{-4x}$ and we assume $y_p = Axe^{4x}$. Substituting into the differential equation we obtain $8A = 2$. Then $A = \frac{1}{4}$, $y_p = \frac{1}{4}xe^{4x}$ and

$$y = c_1 e^{4x} + c_2 e^{-4x} + \frac{1}{4}xe^{4x}.$$

15. From $m^2 + 1 = 0$ we find $m_1 = i$ and $m_2 = -i$. Then $y_c = c_1 \cos x + c_2 \sin x$ and we assume $y_p = (Ax^2 + Bx)\cos x + (Cx^2 + Dx)\sin x$. Substituting into the differential equation we obtain $4C = 0$, $2A + 2D = 0$, $-4A = 2$, and $-2B + 2C = 0$. Then $A = -\frac{1}{2}$, $B = 0$, $C = 0$, $D = \frac{1}{2}$, $y_p = -\frac{1}{2}x^2 \cos x + \frac{1}{2}x \sin x$, and

$$y = c_1 \cos x + c_2 \sin x - \frac{1}{2}x^2 \cos x + \frac{1}{2}x \sin x.$$

18. From $m^2 - 2m + 2 = 0$ we find $m_1 = 1 + i$ and $m_2 = 1 - i$. Then $y_c = e^x(c_1 \cos x + c_2 \sin x)$ and we assume $y_p = Ae^{2x}\cos x + Be^{2x}\sin x$. Substituting into the differential equation we obtain $A + 2B = 1$ and $-2A + B = -3$. Then $A = \frac{7}{5}$, $B = -\frac{1}{5}$, $y_p = \frac{7}{5}e^{2x}\cos x - \frac{1}{5}e^{2x}\sin x$ and

$$y = e^x(c_1 \cos x + c_2 \sin x) + \frac{7}{5}e^{2x}\cos x - \frac{1}{5}e^{2x}\sin x.$$

21. From $m^3 - 6m^2 = 0$ we find $m_1 = m_2 = 0$ and $m_3 = 6$. Then $y_c = c_1 + c_2 x + c_3 e^{6x}$ and we assume $y_p = Ax^2 + B\cos x + C\sin x$. Substituting into the differential equation we obtain $-12A = 3$, $6B - C = -1$, and $B + 6C = 0$. Then $A = -\frac{1}{4}$, $B = -\frac{6}{37}$, $C = \frac{1}{37}$, $y_p = -\frac{1}{4}x^2 - \frac{6}{37}\cos x + \frac{1}{37}\sin x$, and

$$y = c_1 + c_2 x + c_3 e^{6x} - \frac{1}{4}x^2 - \frac{6}{37}\cos x + \frac{1}{37}\sin x.$$

24. From $m^3 - m^2 - 4m + 4 = 0$ we find $m_1 = 1$, $m_2 = 2$, and $m_3 = -2$. Then

$$y_c = c_1 e^x + c_2 e^{2x} + c_3 e^{-2x}$$

and we assume $y_p = A + Bxe^x + Cxe^{2x}$. Substituting into the differential equation we obtain $4A = 5$, $-3B = -1$, and $4C = 1$. Then $A = \frac{5}{4}$, $B = \frac{1}{3}$, $C = \frac{1}{4}$, $y_p = \frac{5}{4} + \frac{1}{3}xe^x + \frac{1}{4}xe^{2x}$, and

$$y = c_1 e^x + c_2 e^{2x} + c_3 e^{-2x} + \frac{5}{4} + \frac{1}{3}xe^x + \frac{1}{4}xe^{2x}.$$

27. We have $y_c = c_1 \cos 2x + c_2 \sin 2x$ and we assume $y_p = A$. Substituting into the differential equation we find $A = -\frac{1}{2}$. Thus $y = c_1 \cos 2x + c_2 \sin 2x - \frac{1}{2}$. From the initial conditions we obtain $c_1 = 0$ and $c_2 = \sqrt{2}$, so $y = \sqrt{2} \sin 2x - \frac{1}{2}$.

30. We have $y_c = c_1 e^{-2x} + c_2 x e^{-2x}$ and we assume $y_p = (Ax^3 + Bx^2)e^{-2x}$. Substituting into the differential equation we find $A = \frac{1}{6}$ and $B = \frac{3}{2}$. Thus $y = c_1 e^{-2x} + c_2 x e^{-2x} + \left(\frac{1}{6}x^3 + \frac{3}{2}x^2\right)e^{-2x}$. From the initial conditions we obtain $c_1 = 2$ and $c_2 = 9$, so

$$y = 2e^{-2x} + 9xe^{-2x} + \left(\frac{1}{6}x^3 + \frac{3}{2}x^2\right)e^{-2x}.$$

33. We have $x_c = c_1 \cos \omega t + c_2 \sin \omega t$ and we assume $x_p = At \cos \omega t + Bt \sin \omega t$. Substituting into the differential equation we find $A = -F_0/2\omega$ and $B = 0$. Thus

$$x = c_1 \cos \omega t + c_2 \sin \omega t - (F_0/2\omega)t \cos \omega t.$$

36. We have $y_c = c_1 e^{-2x} + e^x(c_2 \cos \sqrt{3}\, x + c_3 \sin \sqrt{3}\, x)$ and we assume $y_p = Ax + B + Cxe^{-2x}$. Substituting into the differential equation we find $A = \frac{1}{4}$, $B = -\frac{5}{8}$, and $C = \frac{2}{3}$. Thus

$$y = c_1 e^{-2x} + e^x(c_2 \cos \sqrt{3}\, x + c_3 \sin \sqrt{3}\, x) + \frac{1}{4}x - \frac{5}{8} + \frac{2}{3}xe^{-2x}.$$

From the initial conditions we obtain $c_1 = -\frac{23}{12}$, $c_2 = -\frac{59}{24}$, and $c_3 = \frac{17}{72}\sqrt{3}$, so

$$y = -\frac{23}{12}e^{-2x} + e^x\left(-\frac{59}{24}\cos \sqrt{3}\, x + \frac{17}{72}\sqrt{3}\sin \sqrt{3}\, x\right) + \frac{1}{4}x - \frac{5}{8} + \frac{2}{3}xe^{-2x}.$$

39. The general solution of the differential equation $y'' + 3y = 6x$ is $y = c_1 \cos \sqrt{3}x + c_2 \sin \sqrt{3}x + 2x$. The condition $y(0) = 0$ implies $c_1 = 0$ and so $y = c_2 \sin \sqrt{3}x + 2x$. The condition $y(1) + y'(1) = 0$ implies $c_2 \sin \sqrt{3} + 2 + c_2\sqrt{3}\cos \sqrt{3} + 2 = 0$ so $c_2 = -4/(\sin \sqrt{3} + \sqrt{3}\cos \sqrt{3})$. The solution is

$$y = \frac{-4\sin \sqrt{3}x}{\sin \sqrt{3} + \sqrt{3}\cos \sqrt{3}} + 2x.$$

42. We have $y_c = e^x(c_1 \cos 3x + c_2 \sin 3x)$ and we assume $y_p = A$ on $[0, \pi]$. Substituting into the differential equation we find $A = 2$. Thus, $y = e^x(c_1 \cos 3x + c_2 \sin 3x) + 2$ on $[0, \pi]$. On (π, ∞) we have $y = e^x(c_3 \cos 3x + c_4 \sin 3x)$. From $y(0) = 0$ and $y'(0) = 0$ we obtain

$$c_1 = -2, \qquad c_1 + 3c_2 = 0.$$

Solving this system, we find $c_1 = -2$ and $c_2 = \frac{2}{3}$. Thus $y = e^x(-2\cos 3x + \frac{2}{3}\sin 3x) + 2$ on $[0, \pi]$. Now, continuity of y at $x = \pi$ implies

$$e^\pi\left(-2\cos 3\pi + \frac{2}{3}\sin 3\pi\right) + 2 = e^\pi(c_3 \cos 3\pi + c_4 \sin 3\pi)$$

or $2 + 2e^{\pi} = -c_3 e^{\pi}$ or $c_3 = -2e^{-\pi}(1 + e^{\pi})$. Continuity of y' at π implies

$$\frac{20}{3} e^{\pi} \sin 3\pi = e^{\pi}[(c_3 + 3c_4) \cos 3\pi + (-3c_3 + c_4) \sin 3\pi]$$

or $-c_3 e^{\pi} - 3c_4 e^{\pi} = 0$. Since $c_3 = -2e^{-\pi}(1 + e^{\pi})$ we have $c_4 = \frac{2}{3} e^{-\pi}(1 + e^{\pi})$. The solution of the initial-value problem is

$$y(x) = \begin{cases} e^x(-2\cos 3x + \frac{2}{3}\sin 3x) + 2, & 0 \le x \le \pi \\ (1 + e^{\pi})e^{x-\pi}(-2\cos 3x + \frac{2}{3}\sin 3x), & x > \pi. \end{cases}$$

4.5 Undetermined Coefficients—Annihilator Approach

The terminology and concepts listed below provide an outline of the main ideas encountered in this section. These can be useful when preparing for a quiz or test.

Terminology and Concepts

- annihilator operator

- method of undetermined coefficients

The basic skills listed below summarize the more mechanical types of problems encountered in the exercise set for this section.

Basic Skills

- write a DE using differential operator notation

- find a differential operator that annihilates a given function

- find linearly independent functions that are annihilated by a given differential operator

- solve a nonhomogeneous linear DE with constant coefficients using annihilators and undetermined coefficients

3. $(D^2 - 4D - 12)y = (D - 6)(D + 2)y = x - 6$

6. $(D^3 + 4D)y = D(D^2 + 4)y = e^x \cos 2x$

9. $(D^4 + 8D)y = D(D + 2)(D^2 - 2D + 4)y = 4$

12. $(2D - 1)y = (2D - 1)4e^{x/2} = 8De^{x/2} - 4e^{x/2} = 4e^{x/2} - 4e^{x/2} = 0$

15. D^4 because of x^3

18. $D^2(D - 6)^2$ because of x and xe^{6x}

21. $D^3(D^2 + 16)$ because of x^2 and $\sin 4x$

24. $D(D-1)(D-2)$ because of 1, e^x, and e^{2x}

27. 1, x, x^2, x^3, x^4

30. $D^2 - 9D - 36 = (D-12)(D+3)$; e^{12x}, e^{-3x}

33. $D^3 - 10D^2 + 25D = D(D-5)^2$; 1, e^{5x}, xe^{5x}

36. Applying D to the differential equation we obtain

$$D(2D^2 - 7D + 5)y = 0.$$

Then

$$y = \underbrace{c_1 e^{5x/2} + c_2 e^x}_{y_c} + c_3$$

and $y_p = A$. Substituting y_p into the differential equation yields $5A = -29$ or $A = -29/5$. The general solution is

$$y = c_1 e^{5x/2} + c_2 e^x - \frac{29}{5}.$$

39. Applying D^2 to the differential equation we obtain

$$D^2(D^2 + 4D + 4)y = D^2(D+2)^2 y = 0.$$

Then

$$y = \underbrace{c_1 e^{-2x} + c_2 x e^{-2x}}_{y_c} + c_3 + c_4 x$$

and $y_p = Ax + B$. Substituting y_p into the differential equation yields $4Ax + (4A+4B) = 2x+6$. Equating coefficients gives

$$4A = 2$$

$$4A + 4B = 6.$$

Then $A = 1/2$, $B = 1$, and the general solution is

$$y = c_1 e^{-2x} + c_2 x e^{-2x} + \frac{1}{2}x + 1.$$

42. Applying D^4 to the differential equation we obtain

$$D^4(D^2 - 2D + 1)y = D^4(D-1)^2 y = 0.$$

Then

$$y = \underbrace{c_1 e^x + c_2 x e^x}_{y_c} + c_3 x^3 + c_4 x^2 + c_5 x + c_6$$

and $y_p = Ax^3 + Bx^2 + Cx + E$. Substituting y_p into the differential equation yields

$$Ax \ + (B - 6A)x \ + (6A - 4B + C)x + (2B - 2C + E) = x \ + 4x.$$

Equating coefficients gives

$$A = 1$$
$$B - 6A = 0$$
$$6A - 4B + C = 4$$
$$2B - 2C + E = 0.$$

Then $A = 1$, $B = 6$, $C = 22$, $E = 32$, and the general solution is

$$y = c_1 e^x + c_2 x e^x + x^3 + 6x^2 + 22x + 32.$$

45. Applying $D(D-1)$ to the differential equation we obtain

$$D(D-1)(D^2 - 2D - 3)y = D(D-1)(D+1)(D-3)y = 0.$$

Then

$$y = \underbrace{c_1 e^{3x} + c_2 e^{-x}}_{y_c} + c_3 e^x + c_4$$

and $y_p = Ae^x + B$. Substituting y_p into the differential equation yields $-4Ae^x - 3B = 4e^x - 9$. Equating coefficients gives $A = -1$ and $B = 3$. The general solution is

$$y = c_1 e^{3x} + c_2 e^{-x} - e^x + 3.$$

48. Applying $D(D^2 + 1)$ to the differential equation we obtain

$$D(D^2 + 1)(D^2 + 4)y = 0.$$

Then

$$y = \underbrace{c_1 \cos 2x + c_2 \sin 2x}_{y_c} + c_3 \cos x + c_4 \sin x + c_5$$

and $y_p = A \cos x + B \sin x + C$. Substituting y_p into the differential equation yields

$$3A \cos x + 3B \sin x + 4C = 4 \cos x + 3 \sin x - 8.$$

Equating coefficients gives $A = 4/3$, $B = 1$, and $C = -2$. The general solution is

$$y = c_1 \cos 2x + c_2 \sin 2x + \frac{4}{3} \cos x + \sin x - 2.$$

51. Applying $D(D-1)^3$ to the differential equation we obtain

$$D(D-1)^3(D^2 - 1)y = D(D-1)^4(D+1)y = 0.$$

Then

$$y = \underbrace{c_1 e^x + c_2 e^{-x}}_{y_c} + c_3 x^3 e^x + c_4 x^2 e^x + c_5 x e^x + c_6$$

and $y_p = Ax^3 e^x + Bx^2 e^x + Cxe^x + E$. Substituting y_p into the differential equation yields

$$6Ax^2 e^x \qquad A + 4B \ xe^x \qquad B + 2C \ e^x \qquad E = x^2 e^x + 5$$

Equating coefficients gives

$$6A = 1$$

$$6A + 4B = 0$$

$$2B + 2C = 0$$

$$-E = 5.$$

Then $A = 1/6$, $B = -1/4$, $C = 1/4$, $E = -5$, and the general solution is

$$y = c_1 e^x + c_2 e^{-x} + \frac{1}{6}x^3 e^x - \frac{1}{4}x^2 e^x + \frac{1}{4}xe^x - 5.$$

54. Applying $D^2 - 2D + 10$ to the differential equation we obtain

$$(D^2 - 2D + 10)\left(D^2 + D + \frac{1}{4}\right) y = (D^2 - 2D + 10)\left(D + \frac{1}{2}\right)^2 y = 0.$$

Then

$$y = \underbrace{c_1 e^{-x/2} + c_2 x e^{-x/2}}_{y_c} + c_3 e^x \cos 3x + c_4 e^x \sin 3x$$

and $y_p = Ae^x \cos 3x + Be^x \sin 3x$. Substituting y_p into the differential equation yields

$$(9B - 27A/4)e^x \cos 3x - (9A + 27B/4)e^x \sin 3x = -e^x \cos 3x + e^x \sin 3x.$$

Equating coefficients gives

$$-\frac{27}{4}A + 9B = -1$$

$$-9A - \frac{27}{4}B = 1.$$

Then $A = -4/225$, $B = -28/225$, and the general solution is

$$y = c_1 e^{-x/2} + c_2 x e^{-x/2} - \frac{4}{225}e^x \cos 3x - \frac{28}{225}e^x \sin 3x.$$

57. Applying $(D^2 + 1)^2$ to the differential equation we obtain

$$(D^2 + 1)^2(D^2 + D + 1) = 0.$$

Then

$$y = e^{-x/2}\underbrace{\left[c_1 \cos \frac{\sqrt{3}}{2}x + c_2 \sin \frac{\sqrt{3}}{2}x\right]}_{y_c} + c_3 \cos x + c_4 \sin x + c_5 x \cos x + c_6 x \sin x$$

and $y_p = A \cos x + B \sin x + Cx \cos x + Ex \sin x$. Substituting y_p into the differential equation yields

$$(B + C + 2E) \cos x + Ex \cos x + (-A - 2C + E) \sin x - Cx \sin x = x \sin x.$$

Equating coefficients gives

$$B + C + 2E = 0$$
$$E = 0$$
$$-A - 2C + E = 0$$
$$-C = 1.$$

Then $A = 2$, $B = 1$, $C = -1$, and $E = 0$, and the general solution is

$$y = e^{-x/2}\left[c_1 \cos \frac{\sqrt{3}}{2}x + c_2 \sin \frac{\sqrt{3}}{2}x\right] + 2\cos x + \sin x - x\cos x.$$

60. Applying $D(D-1)^2(D+1)$ to the differential equation we obtain

$$D(D-1)^2(D+1)(D^3 - D^2 + D - 1) = D(D-1)^3(D+1)(D^2+1) = 0.$$

Then

$$y = \underbrace{c_1 e^x + c_2 \cos x + c_3 \sin x}_{y_c} + c_4 + c_5 e^{-x} + c_6 x e^x + c_7 x^2 e^x$$

and $y_p = A + Be^{-x} + Cxe^x + Ex^2 e^x$. Substituting y_p into the differential equation yields

$$4Exe^x + (2C + 4E)e^x - 4Be^{-x} - A = xe^x - e^{-x} + 7.$$

Equating coefficients gives

$$4E = 1$$
$$2C + 4E = 0$$
$$-4B = -1$$
$$-A = 7.$$

Then $A = -7$, $B = 1/4$, $C = -1/2$, and $E = 1/4$, and the general solution is

$$y = c_1 e^x + c_2 \cos x + c_3 \sin x - 7 + \frac{1}{4}e^{-x} - \frac{1}{2}xe^x + \frac{1}{4}x^2 e^x.$$

63. Applying $D(D-1)$ to the differential equation we obtain

$$D(D-1)(D^4 - 2D^3 + D^2) = D^3(D-1)^3 = 0.$$

Then

$$y = \underbrace{c_1 + c_2 x + c_3 e^x + c_4 x e^x}_{y_c} + c_5 x^2 + c_6 x^2 e^x$$

and $y_p = Ax^2 + Bx^2 e^x$. Substituting y_p into the differential equation yields $2A + 2Be^x = 1 + e^x$.
Equating coefficients gives $A = 1/2$ and $B = 1/2$. The general solution is

$$y = c_1 + c_2 x + c_3 e^x + c_4 x e^x + \frac{1}{2}x^2 + \frac{1}{2}x^2 e^x.$$

66. The complementary function is $y_c = c_1 + c_2 e^{-x}$. Using D^2 to annihilate x we find $y_p = Ax + Bx^2$. Substituting y_p into the differential equation we obtain $(A + 2B) + 2Bx = x$. Thus $A = -1$ and $B = 1/2$, and

$$y = c_1 + c_2 e^{-x} - x + \frac{1}{2}x^2$$
$$y' = -c_2 e^{-x} - 1 + x.$$

The initial conditions imply

$$c_1 + c_2 = 1$$
$$-c_2 = 1.$$

Thus $c_1 = 2$ and $c_2 = -1$, and

$$y = 2 - e^{-x} - x + \frac{1}{2}x^2.$$

69. The complementary function is $y_c = c_1 \cos x + c_2 \sin x$. Using $(D^2 + 1)(D^2 + 4)$ to annihilate $8 \cos 2x - 4 \sin x$ we find $y_p = Ax \cos x + Bx \sin x + C \cos 2x + E \sin 2x$. Substituting y_p into the differential equation we obtain $2B \cos x - 3C \cos 2x - 2A \sin x - 3E \sin 2x = 8 \cos 2x - 4 \sin x$. Thus $A = 2$, $B = 0$, $C = -8/3$, and $E = 0$, and

$$y = c_1 \cos x + c_2 \sin x + 2x \cos x - \frac{8}{3} \cos 2x$$
$$y' = -c_1 \sin x + c_2 \cos x + 2 \cos x - 2x \sin x + \frac{16}{3} \sin 2x.$$

The initial conditions imply

$$c_2 + \frac{8}{3} = -1$$
$$-c_1 - \pi = 0.$$

Thus $c_1 = -\pi$ and $c_2 = -11/3$, and

$$y = -\pi \cos x - \frac{11}{3} \sin x + 2x \cos x - \frac{8}{3} \cos 2x.$$

72. The complementary function is $y_c = c_1 + c_2 x + c_3 x^2 + c_4 e^x$. Using $D^2(D - 1)$ to annihilate $x + e^x$ we find $y_p = Ax^3 + Bx^4 + Cxe^x$. Substituting y_p into the differential equation we obtain

$$(-6A + 24B) - 24Bx + Ce^x = x + e^x. \text{ Thus } A = -1/6, \ B = -1/24, \text{ and } C = 1, \text{ and}$$

$$y = c_1 + c_2 x + c_3 x^2 + c_4 e^x - \frac{1}{6} x^3 - \frac{1}{24} x^4 + x e^x$$

$$y' = c_2 + 2 c_3 x + c_4 e^x - \frac{1}{2} x^2 - \frac{1}{6} x^3 + e^x + x e^x$$

$$y'' = 2 c_3 + c_4 e^x - x - \frac{1}{2} x^2 + 2 e^x + x e^x.$$

$$y''' = c_4 e^x - 1 - x + 3 e^x + x e^x.$$

The initial conditions imply

$$c_1 + c_4 = 0$$

$$c_2 + c_4 + 1 = 0$$

$$2 c_3 + c_4 + 2 = 0$$

$$2 + c_4 = 0.$$

Thus $c_1 = 2$, $c_2 = 1$, $c_3 = 0$, and $c_4 = -2$, and

$$y = 2 + x - 2 e^x - \frac{1}{6} x^3 - \frac{1}{24} x^4 + x e^x.$$

4.6   Variation of Parameters

The terminology and concepts listed below provide an outline of the main ideas encountered in this section. These can be useful when preparing for a quiz or test.

Terminology and Concepts

- variation of parameters

The basic skills listed below summarize the more mechanical types of problems encountered in the exercise set for this section.

Basic Skills

- solve a nonhomogeneous linear second-order DE with constant coefficients using variation of parameters

Integrals of Trig Functions

In Example 2 in the text we had to integrate

$$\int \cot 3x \, dx = \int \frac{\cos 3x}{\sin 3x} \, dx.$$

If we identify $u = \sin 3x$, then $du = 3 \cos 3x \, dx$. The integral is recognized as one of the form $\int du/u = \ln|u| + c$:

$$\frac{1}{3} \int \frac{\overbrace{3 \cos 3x \, dx}^{du}}{\underbrace{\sin 3x}_{u}} = \frac{1}{3} \ln|\sin 3x| + c.$$

In general, we can integrate $\tan u$ and $\cot u$ in terms of logarithms:

$$\int \tan u \, du = -\ln|\cos u| + c \tag{1}$$

$$\int \cot u \, du = \ln|\sin u| + c. \tag{2}$$

The integrals in (1) and (2), along with

$$\int \sec u \, du = \ln|\sec u + \tan u| + c \tag{3}$$

$$\int \csc u \, du = \ln|\csc u - \cot u| + c \tag{4}$$

occur frequently in the solution of differential equations by variation of parameters. Watch for them in Exercises 4.6.

The particular solution, $y_p = u_1 y_1 + u_2 y_2$, in the following problems can take on a variety of forms, especially where trigonometric functions are involved. The validity of a particular form can best be checked by substituting it back into the differential equation.

In the remaining problems in this section the value of the Wronskian is given, but the determinant which is used to evaluate the Wronskian is not shown.

3. The auxiliary equation is $m^2 + 1 = 0$, so $y_c = c_1 \cos x + c_2 \sin x$ and $W = 1$. Identifying $f(x) = \sin x$ we obtain

$$u_1' = -\sin^2 x$$
$$u_2' = \cos x \sin x.$$

Then

$$u_1 = \frac{1}{4} \sin 2x - \frac{1}{2}x = \frac{1}{2} \sin x \cos x - \frac{1}{2}x$$

$$u_2 = -\frac{1}{2} \cos^2 x.$$

and

$$y = c_1 \cos x + c_2 \sin x + \frac{1}{2} \sin x \cos^2 x - \frac{1}{2}x \cos x - \frac{1}{2} \cos^2 x \sin x$$

$$= c_1 \cos x + c_2 \sin x - \frac{1}{2}x \cos x.$$

6. The auxiliary equation is $m^2 + 1 = 0$, so $y_c = c_1 \cos x + c_2 \sin x$ and $W = 1$. Identifying $f(x) = \sec^2 x$ we obtain

$$u_1' = -\frac{\sin x}{\cos^2 x}$$

$$u_2' = \sec x.$$

Then

$$u_1 = -\frac{1}{\cos x} = -\sec x$$

$$u_2 = \ln|\sec x + \tan x|$$

and

$$y = c_1 \cos x + c_2 \sin x - \cos x \sec x + \sin x \ln|\sec x + \tan x|$$

$$= c_1 \cos x + c_2 \sin x - 1 + \sin x \ln|\sec x + \tan x|.$$

9. The auxiliary equation is $m^2 - 4 = 0$, so $y_c = c_1 e^{2x} + c_2 e^{-2x}$ and $W = -4$. Identifying $f(x) = e^{2x}/x$ we obtain $u_1' = 1/4x$ and $u_2' = -e^{4x}/4x$. Then

$$u_1 = \frac{1}{4}\ln|x|,$$

$$u_2 = -\frac{1}{4}\int_{x_0}^{x}\frac{e^{4t}}{t}\,dt$$

and

$$y = c_1 e^{2x} + c_2 e^{-2x} + \frac{1}{4}\left(e^{2x}\ln|x| - e^{-2x}\int_{x_0}^{x}\frac{e^{4t}}{t}\,dt\right), \qquad x_0 > 0.$$

12. The auxiliary equation is $m^2 - 2m + 1 = (m-1)^2 = 0$, so $y_c = c_1 e^x + c_2 x e^x$ and $W = e^{2x}$. Identifying $f(x) = e^x/(1+x^2)$ we obtain

$$u_1' = -\frac{xe^x e^x}{e^{2x}(1+x^2)} = -\frac{x}{1+x^2}$$

$$u_2' = \frac{e^x e^x}{e^{2x}(1+x^2)} = \frac{1}{1+x^2}.$$

Then $u_1 = -\frac{1}{2}\ln(1+x^2)$, $u_2 = \tan^{-1} x$, and

$$y = c_1 e^x + c_2 x e^x - \frac{1}{2}e^x\ln(1+x^2) + xe^x\tan^{-1} x.$$

15. The auxiliary equation is $m^2 + 2m + 1 = (m+1)^2 = 0$, so $y_c = c_1 e^{-t} + c_2 t e^{-t}$ and $W = e^{-2t}$. Identifying $f(t) = e^{-t} \ln t$ we obtain

$$u_1' = -\frac{t e^{-t} e^{-t} \ln t}{e^{-2t}} = -t \ln t$$

$$u_2' = \frac{e^{-t} e^{-t} \ln t}{e^{-2t}} = \ln t.$$

Then

$$u_1 = -\frac{1}{2} t^2 \ln t + \frac{1}{4} t^2$$

$$u_2 = t \ln t - t$$

and

$$y = c_1 e^{-t} + c_2 t e^{-t} - \frac{1}{2} t^2 e^{-t} \ln t + \frac{1}{4} t^2 e^{-t} + t^2 e^{-t} \ln t - t^2 e^{-t}$$

$$= c_1 e^{-t} + c_2 t e^{-t} + \frac{1}{2} t^2 e^{-t} \ln t - \frac{3}{4} t^2 e^{-t}.$$

18. The auxiliary equation is $4m^2 - 4m + 1 = (2m-1)^2 = 0$, so $y_c = c_1 e^{x/2} + c_2 x e^{x/2}$ and $W = e^x$. Identifying $f(x) = \frac{1}{4} e^{x/2} \sqrt{1 - x^2}$ we obtain

$$u_1' = -\frac{x e^{x/2} e^{x/2} \sqrt{1-x^2}}{4 e^x} = -\frac{1}{4} x \sqrt{1 - x^2}$$

$$u_2' = \frac{e^{x/2} e^{x/2} \sqrt{1-x^2}}{4 e^x} = \frac{1}{4} \sqrt{1 - x^2}.$$

To find u_1 and u_2 we use the substitution $v = 1 - x^2$ and the trig substitution $x = \sin \theta$, respectively:

$$u_1 = \frac{1}{12} \left(1 - x^2\right)^{3/2}$$

$$u_2 = \frac{x}{8} \sqrt{1 - x^2} + \frac{1}{8} \sin^{-1} x.$$

Thus

$$y = c_1 e^{x/2} + c_2 x e^{x/2} + \frac{1}{12} e^{x/2} \left(1 - x^2\right)^{3/2} + \frac{1}{8} x^2 e^{x/2} \sqrt{1 - x^2} + \frac{1}{8} x e^{x/2} \sin^{-1} x.$$

21. The auxiliary equation is $m^2 + 2m - 8 = (m - 2)(m + 4) = 0$, so $y_c = c_1 e^{2x} + c_2 e^{-4x}$ and $W = -6e^{-2x}$. Identifying $f(x) = 2e^{-2x} - e^{-x}$ we obtain

$$u_1' = \frac{1}{3}e^{-4x} - \frac{1}{6}e^{-3x}$$

$$u_2' = \frac{1}{6}e^{3x} - \frac{1}{3}e^{2x}.$$

Then

$$u_1 = -\frac{1}{12}e^{-4x} + \frac{1}{18}e^{-3x}$$

$$u_2 = \frac{1}{18}e^{3x} - \frac{1}{6}e^{2x}.$$

Thus

$$y = c_1 e^{2x} + c_2 e^{-4x} - \frac{1}{12}e^{-2x} + \frac{1}{18}e^{-x} + \frac{1}{18}e^{-x} - \frac{1}{6}e^{-2x}$$

$$= c_1 e^{2x} + c_2 e^{-4x} - \frac{1}{4}e^{-2x} + \frac{1}{9}e^{-x}$$

and

$$y' = 2c_1 e^{2x} - 4c_2 e^{-4x} + \frac{1}{2}e^{-2x} - \frac{1}{9}e^{-x}.$$

The initial conditions imply

$$c_1 + c_2 - \frac{5}{36} = 1$$

$$2c_1 - 4c_2 + \frac{7}{18} = 0.$$

Thus $c_1 = 25/36$ and $c_2 = 4/9$, and

$$y = \frac{25}{36}e^{2x} + \frac{4}{9}e^{-4x} - \frac{1}{4}e^{-2x} + \frac{1}{9}e^{-x}.$$

24. Write the equation in the form

$$y'' + \frac{1}{x}y' + \frac{1}{x^2}y = \frac{\sec(\ln x)}{x^2}$$

and identify $f(x) = \sec(\ln x)/x^2$. From $y_1 = \cos(\ln x)$ and $y_2 = \sin(\ln x)$ we compute $W = 1/x$.

Now

$$u_1' = -\frac{\tan(\ln x)}{x} \quad \text{so} \quad u_1 = \ln|\cos(\ln x)|$$

and

$$u_2' = \frac{1}{x} \quad \text{so} \quad u_2 = \ln x.$$

Thus, a particular solution is

$$y_p = \cos(\ln x) \ln |\cos(\ln x)| + (\ln x) \sin(\ln x),$$

and the general solution is

$$y = c_1 \cos(\ln x) + c_2 \sin(\ln x) + \cos(\ln x) \ln |\cos(\ln x)| + (\ln x) \sin(\ln x).$$

27. The auxiliary equation is $m^3 - 2m^2 - m + 2 = 0$ so $y_c = c_1 e^x + c_2 e^{-x} + c_3 e^{2x}$ and $W = -6e^{2x}$. Identifying $f(x) = e^{4x}$ we find $W_1 = 3e^{5x}$, $W_2 = -e^{7x}$, and $W_3 = -2e^{4x}$. Then

$$u_1' = \frac{W_1}{W} = \frac{3e^{5x}}{-6e^{2x}} = -\frac{1}{2}e^{3x}$$

$$u_2' = \frac{W_2}{W} = \frac{-e^{7x}}{-6e^{2x}} = \frac{1}{6}e^{5x}$$

$$u_3' = \frac{W_3}{W} = \frac{-2e^{4x}}{-6e^{2x}} = \frac{1}{3}e^{2x},$$

and integrating we find that

$$u_1 = -\frac{1}{6}e^{3x}$$

$$u_2 = \frac{1}{30}e^{5x}$$

$$u_3 = \frac{1}{6}e^{2x}.$$

Thus $y_p = -\frac{1}{6}e^{3x} \cdot e^x + \frac{1}{30}e^{5x} \cdot e^{-x} + \frac{1}{6}e^{2x} \cdot e^{2x} = \frac{1}{30}e^{4x}$, and the general solution is

$$y = c_1 e^x + c_2 e^{-x} + c_3 e^{2x} + \frac{1}{30}e^{4x}.$$

4.7 | Cauchy-Euler Equation

The terminology and concepts listed below provide an outline of the main ideas encountered in this section. These can be useful when preparing for a quiz or test.

Terminology and Concepts

- Cauchy-Euler equation
- auxiliary equation
- variation of parameters

The basic skills listed below summarize the more mechanical types of problems encountered in the exercise set for this section.

Basic Skills

- find the general solution of a homogeneous Cauchy-Euler equation
- use variation of parameters to find a particular solution of a nonhomogeneous Cauchy-Euler equation
- use the substitution $x = e^t$ to transform a Cauchy-Euler equation into a linear DE with constant coefficients

3. The auxiliary equation is $m^2 = 0$ so that $y = c_1 + c_2 \ln x$.

6. The auxiliary equation is $m^2 + 4m + 3 = (m+1)(m+3) = 0$ so that $y = c_1 x^{-1} + c_2 x^{-3}$.

9. The auxiliary equation is $25m^2 + 1 = 0$ so that $y = c_1 \cos\left(\frac{1}{5}\ln x\right) + c_2 \sin\left(\frac{1}{5}\ln x\right)$.

12. The auxiliary equation is $m^2 + 7m + 6 = (m+1)(m+6) = 0$ so that $y = c_1 x^{-1} + c_2 x^{-6}$.

15. Assuming that $y = x^m$ and substituting into the differential equation we obtain

$$m(m-1)(m-2) - 6 = m^3 - 3m^2 + 2m - 6 = (m-3)(m^2+2) = 0.$$

Thus

$$y = c_1 x^3 + c_2 \cos\left(\sqrt{2}\ln x\right) + c_3 \sin\left(\sqrt{2}\ln x\right).$$

18. Assuming that $y = x^m$ and substituting into the differential equation we obtain

$$m(m-1)(m-2)(m-3) + 6m(m-1)(m-2) + 9m(m-1) + 3m + 1 = m^4 + 2m^2 + 1$$
$$= (m^2+1)^2 = 0.$$

Thus

$$y = c_1 \cos(\ln x) + c_2 \sin(\ln x) + c_3(\ln x)\cos(\ln x) + c_4(\ln x)\sin(\ln x).$$

21. The auxiliary equation is $m^2 - 2m + 1 = (m-1)^2 = 0$ so that $y_c = c_1 x + c_2 x \ln x$ and the Wronskian is $W(x, x \ln x) = x$. Identifying $f(x) = 2/x$ we obtain $u_1' = -2 \ln x / x$ and $u_2' = 2/x$. Then $u_1 = -(\ln x)^2$, $u_2 = 2 \ln x$, and

$$y = c_1 x + c_2 x \ln x - x(\ln x)^2 + 2x(\ln x)^2$$
$$= c_1 x + c_2 x \ln x + x(\ln x)^2, \qquad x > 0.$$

24. The auxiliary equation $m(m-1) + m - 1 = m^2 - 1 = 0$ has roots $m_1 = -1$, $m_2 = 1$, so $y_c = c_1 x^{-1} + c_2 x$. With $y_1 = x^{-1}$, $y_2 = x$, and the identification $f(x) = 1/x^2(x+1)$, we get

$$W = 2x^{-1}, \qquad W_1 = -1/x(x+1), \qquad \text{and} \qquad W_2 = 1/x^3(x+1).$$

Then $u_1' = W_1/W = -1/2(x+1)$, $u_2' = W_2/W = 1/2x^2(x+1)$, and integration by partial fractions for u_2' gives

$$u_1 = -\frac{1}{2}\ln(x+1)$$
$$u_2 = -\frac{1}{2}x^{-1} - \frac{1}{2}\ln x + \frac{1}{2}\ln(x+1),$$

so

$$y_p = u_1 y_1 + u_2 y_2 = \left[-\frac{1}{2}\ln(x+1)\right]x^{-1} + \left[-\frac{1}{2}x^{-1} - \frac{1}{2}\ln x + \frac{1}{2}\ln(x+1)\right]x$$
$$= -\frac{1}{2} - \frac{1}{2}x\ln x + \frac{1}{2}x\ln(x+1) - \frac{\ln(x+1)}{2x} = -\frac{1}{2} + \frac{1}{2}x\ln\left(1 + \frac{1}{x}\right) - \frac{\ln(x+1)}{2x}$$

and

$$y = y_c + y_p = c_1 x^{-1} + c_2 x - \frac{1}{2} + \frac{1}{2}x\ln\left(1 + \frac{1}{x}\right) - \frac{\ln(x+1)}{2x}, \qquad x > 0.$$

27. The auxiliary equation is $m^2 + 1 = 0$, so that

$$y = c_1 \cos(\ln x) + c_2 \sin(\ln x)$$

and

$$y' = -c_1 \frac{1}{x}\sin(\ln x) + c_2 \frac{1}{x}\cos(\ln x).$$

The initial conditions imply $c_1 = 1$ and $c_2 = 2$. Thus $y = \cos(\ln x) + 2\sin(\ln x)$. The graph is given to the right.

30. The auxiliary equation is

$$m^2 - 6m + 8 = (m-2)(m-4) = 0,$$

so that $y_c = c_1 x^2 + c_2 x^4$ and the Wronskian is $W = 2x^5$. Identifying $f(x) = 8x^4$ we obtain $u_1' = -4x^3$ and $u_2' = 4x$. Then $u_1 = -x^4$, $u_2 = 2x^2$, and $y = c_1 x^2 + c_2 x^4 + x^6$. The initial conditions imply

$$\frac{1}{4}c_1 + \frac{1}{16}c_2 = -\frac{1}{64}$$
$$c_1 + \frac{1}{2}c_2 = -\frac{3}{16}.$$

Thus $c_1 = \frac{1}{16}$, $c_2 = -\frac{1}{2}$, and $y = \frac{1}{16}x^2 - \frac{1}{2}x^4 + x^6$. The graph is given above.

33. Substituting $x = e^t$ into the differential equation we obtain

$$\frac{d^2 y}{dt^2} + 9\frac{dy}{dt} + 8y = e^{2t}.$$

The auxiliary equation is $m^2 + 9m + 8 = (m+1)(m+8) = 0$ so that $y_c = c_1 e^{-t} + c_2 e^{-8t}$. Using undetermined coefficients we try $y_p = Ae^{2t}$. This leads to $30Ae^{2t} = e^{2t}$, so that $A = 1/30$ and

$$y = c_1 e^{-t} + c_2 e^{-8t} + \frac{1}{30}e^{2t} = c_1 x^{-1} + c_2 x^{-8} + \frac{1}{30}x^2.$$

36. From

$$\frac{d^2 y}{dx^2} = \frac{1}{x^2}\left(\frac{d^2 y}{dt^2} - \frac{dy}{dt}\right)$$

it follows that

$$\frac{d^3 y}{dx^3} = \frac{1}{x^2}\frac{d}{dx}\left(\frac{d^2 y}{dt^2} - \frac{dy}{dt}\right) - \frac{2}{x^3}\left(\frac{d^2 y}{dt^2} - \frac{dy}{dt}\right)$$

$$= \frac{1}{x^2}\frac{d}{dx}\left(\frac{d^2 y}{dt^2}\right) - \frac{1}{x^2}\frac{d}{dx}\left(\frac{dy}{dt}\right) - \frac{2}{x^3}\frac{d^2 y}{dt^2} + \frac{2}{x^3}\frac{dy}{dt}$$

$$= \frac{1}{x^2}\frac{d^3 y}{dt^3}\left(\frac{1}{x}\right) - \frac{1}{x^2}\frac{d^2 y}{dt^2}\left(\frac{1}{x}\right) - \frac{2}{x^3}\frac{d^2 y}{dt^2} + \frac{2}{x^3}\frac{dy}{dt}$$

$$= \frac{1}{x^3}\left(\frac{d^3 y}{dt^3} - 3\frac{d^2 y}{dt^2} + 2\frac{dy}{dt}\right).$$

Substituting into the differential equation we obtain

$$\frac{d^3y}{dt^3} - 3\frac{d^2y}{dt^2} + 2\frac{dy}{dt} - 3\left(\frac{d^2y}{dt^2} - \frac{dy}{dt}\right) + 6\frac{dy}{dt} - 6y = 3 + 3t$$

or

$$\frac{d^3y}{dt^3} - 6\frac{d^2y}{dt^2} + 11\frac{dy}{dt} - 6y = 3 + 3t.$$

The auxiliary equation is $m^3 - 6m^2 + 11m - 6 = (m-1)(m-2)(m-3) = 0$ so that $y_c = c_1e^t + c_2e^{2t} + c_3e^{3t}$. Using undetermined coefficients we try $y_p = A + Bt$. This leads to $(11B - 6A) - 6Bt = 3 + 3t$, so that $A = -17/12$, $B = -1/2$, and

$$y = c_1e^t + c_2e^{2t} + c_3e^{3t} - \frac{17}{12} - \frac{1}{2}t = c_1x + c_2x^2 + c_3x^3 - \frac{17}{12} - \frac{1}{2}\ln x.$$

39. Letting $y = (x+3)^m$ we have $y' = m(x+3)^{m-1}$, and $y'' = m(m-1)(x+3)^{m-2}$. Substituting into the differential equation we find

$$(x+3)^2 m(m-1)(x+3)^{m-2} - 8(x+3)m(x+3)^{m-1} + 14(x+3)^m$$

$$= m(m-1)(x+3)^m - 8m(x+3)^m + 14(x+3)^m = 0,$$

so $m(m-1) - 8m + 14 = m^2 - 9m + 14 = (m-2)(m-7) = 0$. Thus, $m_1 = 2$ and $m_2 = 7$, and the general solution is

$$y = c_1(x+3)^2 + c_2(x+3)^7.$$

42. Letting $t = x - 4$ we obtain $dy/dx = dy/dt$ and, using the Chain Rule,

$$\frac{d^2y}{dx^2} = \frac{d}{dx}\left(\frac{dy}{dt}\right) = \frac{d^2y}{dt^2}\frac{dt}{dx} = \frac{d^2y}{dt^2}(1) = \frac{d^2y}{dt^2}.$$

Substituting into the differential equation we obtain

$$t^2\frac{d^2y}{dt^2} - 5t\frac{dy}{dt} + 9y = 0.$$

The auxiliary equation is $m(m-1) - 5m + 9 = m^2 - 6m + 9 = (m-3)^2 = 0$ so that $m_1 = m_2 = 3$. Then

$$y = c_1t^3 + c_2t^3\ln t = c_1(x-4)^3 + c_2(x-4)^3\ln(x-4).$$

4.8 | Green's Functions

The terminology and concepts listed below provide an outline of the main ideas encountered in this section. These can be useful when preparing for a quiz or test.

Terminology and Concepts

- rest solution

- Green's function (or Green's function for the second-order differential operator)

- response of system due to initial conditions

- response of system due to the forcing function f

- Green's function for a boundary-value problem

Basic Skills

- use Green's functions to solve initial-value problems

- use Green's functions to solve initial-value problems with piecewise-defined forcing functions

- use Green's functions to solve boundary-value problems

4.8.1 Initial-Value Problems

---

3. From the homogeneous differential equation $y'' + 2y' + y = 0$, we find $y_1 = e^{-x}$ and $y_2 = xe^{-x}$ so

$$W\left(e^{-x},\, xe^{-x}\right) = \begin{vmatrix} e^{-x} & xe^{-x} \\ -e^{-x} & -xe^{-x} + e^{-x} \end{vmatrix} = e^{-2x}$$

Then

$$G(x,\, t) = \frac{e^{-t}xe^{-x} - e^{-x}te^{-t}}{e^{-2t}} = (x - t)e^{-(x-t)}$$

and

$$y_p(x) = \int_{x_0}^{x} (x - t)e^{-(x-t)} f(t)\, dt.$$

6. From the homogeneous differential equation $y'' - 2y' + 2y = 0$, we find $y_1 = e^x \cos x$ and $y_2 = e^x \sin x$ so

$$W(e^x \cos x, e^x \sin x) = \begin{vmatrix} e^x \cos x & e^x \sin x \\ -e^x \sin x + e^x \cos x & e^x \cos x + e^x \sin x \end{vmatrix} = e^{2x}$$

Then

$$G(x, t) = \frac{e^t \cos t\, e^x \sin x - e^x \cos x\, e^t \sin t}{e^{2t}} = e^{x-t} \sin(x - t)$$

and

$$y_p(x) = \int_{x_0}^x e^{x-t} \sin(x - t) f(t)\, dt.$$

9. From $y'' + 2y' + y = e^{-x}$ we have

$$y = y_c + y_p = c_1 e^{-x} + c_2 x e^{-x} + y_p,$$

so

$$y = c_1 e^{-x} + c_2 x e^{-x} + \int_{x_0}^x (x - t) e^{-(x-t)} e^{-t}\, dt.$$

12. From $y'' - 2y' + 2y = \cos^2 x$ we have

$$y = y_c + y_p = c_1 e^x \cos x + c_2 e^x \sin x + y_p,$$

so

$$y = c_1 e^x \cos x + c_2 e^x \sin x = \int_{x_0}^x e^{(x-t)} \sin(x - t) \cos^2 t\, dt.$$

15. The initial-value problem is $y'' - 10y' + 25y = e^{5x}$, $\quad y(0) = 0$, $y'(0) = 0$. Then we find that

$$y_1 = e^{5x},\ y_2 = xe^{5x} \quad \text{so} \quad W(e^{5x}, xe^{5x}) = \begin{vmatrix} e^{5x} & xe^{5x} \\ 5e^{5x} & 5xe^{5x} + e^{5x} \end{vmatrix} = e^{10x}.$$

Then $G(x, t) = \dfrac{e^{5t} x e^{5x} - e^{5x} t e^{5t}}{e^{10t}} = (x - t) e^{5(x-t)}$ and the solution of the initial-value problem is

$$y_p(x) = \int_0^x (x - t) e^{5(x-t)} e^{5t}\, dt = xe^{5x} \int_0^x dt - e^{5x} \int_0^x t\, dt = x^2 e^{5x} - \frac{1}{2} x^2 e^{5x} = \frac{1}{2} x^2 e^{5x}.$$

18. The initial-value problem is $y'' + y = \sec^2 x$, $y(\pi) = 0$, $y'(\pi) = 0$. Then we find that

$$y_1 = \cos x, \ y_2 = \sin x \qquad \text{so} \qquad W(\cos x, \sin x) = \begin{vmatrix} \cos x & \sin x \\ -\sin x & \cos x \end{vmatrix} = 1.$$

Then $G(x, t) = \cos t \sin x - \cos x \sin t = \sin(x - t)$ and the solution of the initial-value problem is

$$y_p(x) = \int_\pi^x (\cos t \sin x - \cos x \sin t) \sec^2 t \, dt$$

$$= \sin x \int_\pi^x \sec t \, dt - \cos x \int_\pi^x \sec t \tan t \, dt$$

$$= -\cos x - 1 + \sin x \ln|\sec x + \tan x|.$$

21. The initial-value problem is $y'' - 10y' + 25y = e^{5x}$, $y(0) = -1$, $y'(0) = 1$, so $y(x) = c_1 e^{5x} + c_2 x e^{5x}$. The initial conditions give

$$c_1 \qquad = -1$$
$$5c_1 + c_2 = \quad 1,$$

so $c_1 = -1$ and $c_2 = 6$, which implies that $y_h = -e^{5x} + 6xe^{5x}$. Now, y_p found in the solution of Problem 15 in this section gives

$$y = y_h + y_p = -e^{5x} + 6xe^{5x} + \frac{1}{2}x^2 e^{5x}.$$

24. The initial-value problem is $y'' + y = \sec^2 x$, $y(\pi) = \frac{1}{2}$, $y'(\pi) = -1$, so $y(x) = c_1 \cos x + c_2 \sin x$. The initial conditions give

$$-c_1 \qquad = \frac{1}{2}$$
$$-c_2 = -1,$$

so $c_1 = -\frac{1}{2}$ and $c_2 = 1$, which implies that $y_h = -\frac{1}{2}\cos x + \sin x$. Now, y_p found in the solution of Problem 18 in this section gives

$$y = y_h + y_p = -\frac{1}{2}\cos x + \sin x + (-\cos x - 1 + \sin x \ln|\sec x + \tan x|)$$

$$= -\frac{3}{2}\cos x + \sin x - 1 + \sin x \ln|\sec x + \tan x|.$$

27. The Cauchy-Euler initial-value problem $x^2y'' - 2xy' + 2y = x$, $y(1) = 2$, $y'(1) = -1$, has

auxiliary equation $m(m-1) - 2m + 2 = m^2 - 3m + 2 = (m-1)(m-2) = 0$ so $m_1 = 1$, $m_2 = 2$,

$y(x) = c_1 x + c_2 x^2$, and $y' = c_1 + 2c_2 x$. The initial conditions give

$$c_1 + c_2 = 2$$
$$c_1 + 2c_2 = -1,$$

so $c_1 = 5$ and $c_2 = -3$, which implies that $y_h = 5x - 3x^2$. The Wronskian is

$$W\left(x, x^2\right) = \begin{vmatrix} x & x^2 \\ 1 & 2x \end{vmatrix} = x^2.$$

Then $G(x, t) = \dfrac{tx^2 - xt^2}{t^2} = \dfrac{x(x-t)}{t}$. From the standard form of the differential equation we

identify the forcing function $f(t) = \dfrac{1}{t}$. Then, for $x > 1$,

$$y_p(x) = \int_1^x \frac{x(x-t)}{t} \frac{1}{t} \, dt = x^2 \int_1^x \frac{1}{t^2} \, dt - x \int_1^x \frac{1}{t} \, dt$$

$$= x^2 \left(-\frac{1}{x} + 1\right) - x(\ln x - \ln 1) = x^2 - x - x \ln x.$$

The solution of the initial-value problem is

$$y = y_h + y_p = (5x - 3x^2) + (x^2 - x - x \ln x) = 4x - 2x^2 - x \ln x.$$

30. The Cauchy-Euler initial-value problem $x^2y'' - xy' + y = x^2$, $y(1) = 4$, $y'(1) = 3$, has

auxiliary equation $m(m-1) - m + 1 = m^2 - 2m + 1 = (m-1)^2 = 0$ so $m_1 = m_2 = 1$,

$y(x) = c_1 x + c_2 x \ln x$, and $y' = c_1 + c_2(1 + \ln x)$. The initial conditions give

$$c_1 = 4$$
$$c_1 + c_2 = 3,$$

so $c_1 = 4$ and $c_2 = -1$, which implies that $y_h = 4x - x \ln x$. The Wronskian is

$$W(x, x \ln x) = \begin{vmatrix} x & x \ln x \\ 1 & 1 + \ln x \end{vmatrix} = x.$$

Then $G(x, t) = \dfrac{tx \ln x - xt \ln t}{t} = x \ln x - x \ln t$. From the standard form of the differential equation we identify the forcing function $f(t) = 1$. Then, for $x > 1$,

$$y_p(x) = \int_1^x (x \ln x - x \ln t)\, dt = x \ln x \int_1^x dt - x \int_1^x \ln t\, dt$$

$$= x \ln x\,(x - 1) - x(x \ln x - x + 1) = x^2 - x - x \ln x.$$

The solution of the initial-value problem is

$$y = y_h + y_p = 4x - x \ln x + \left(x^2 - x - x \ln x\right) = x^2 + 3x - 2x \ln x.$$

33. The initial-value problem is $y'' + y = f(x), \quad y(0) = 1,\ y'(0) = -1$, where

$$f(x) = \begin{cases} 0, & x < 0 \\ 10, & 0 \le x \le 3\pi \\ 0, & x > 3\pi. \end{cases}$$

We first find

$$y_h(x) = \cos x - \sin x \qquad \text{and} \qquad y_p(x) = \int_0^x \sin(x - t)\, f(t)\, dt.$$

Then for $x < 0$

$$y_p(x) = \int_0^x \sin(x - t)\, 0\, dt = 0,$$

for $0 \le x \le 3\pi$

$$y_p(x) = 10 \int_0^x \sin(x - t)\, dt = 10 - 10 \cos x,$$

and for $x > 3\pi$

$$y_p(x) = 10 \int_0^{3\pi} \sin(x - t)\, dt + \int_{3\pi}^x \sin(x - t)\, 0\, dt = -20 \cos x.$$

The solution is

$$y(x) = y_h(x) + y_p(x) = \cos x - \sin x + y_p(x),$$

where

$$y_p(x) = \begin{cases} 0, & x < 0 \\ 10 - 10 \cos x, & 0 \le x \le 3\pi \\ -20 \cos x, & x > 3\pi. \end{cases}$$

4.8.2 Boundary-Value Problems

36. The boundary-value problem is $y'' = f(x)$, $y(0) = 0$, $y(1) + y'(1) = 0$. The solution of the associated homogeneous equation is $y = c_1 + c_2 x$.

(a) To satisfy $y(0) = 0$ we take $y_1(x) = x$ and to satisfy $y(1) + y'(1) = 0$ we note that $y(1) + y'(1) = (c_1 + c_2) + c_2 = 0$ which implies that $c_1 = -2c_2$. Taking $c_2 = 1$, we find that $c_1 = -2$, so we have $y_2(x) = -2 + x$. The Wronskian of y_1 and y_2 is

$$W(y_1, y_2) = \begin{vmatrix} x & -2 + x \\ 1 & 1 \end{vmatrix} = 2,$$

so

$$G(x, t) = \begin{cases} \frac{1}{2} t(x - 2), & 0 \le t \le x \\ \frac{1}{2} x(t - 2), & x \le t \le 1. \end{cases}$$

Therefore

$$y_p(x) = \int_0^1 G(x, t) \, f(t) \, dt = \frac{1}{2}(x - 2) \int_0^x t f(t) \, dt + \frac{1}{2} x \int_x^1 (t - 2) f(t) \, dt.$$

(b) By the Product Rule and the Fundamental Theorem of Calculus, the first two derivative of $y_p(x)$ are

$$y_p'(x) = \frac{1}{2}(x - 2) x f(x) + \frac{1}{2} \int_0^x t f(t) \, dt + \frac{1}{2} x[-(x - 2) f(x)] + \frac{1}{2} \int_x^1 (t - 2) f(t) \, dt$$

$$y_p''(x) = \frac{1}{2}(x - 2)[x f'(x) + f(x)] + \frac{1}{2} x f(x) + \frac{1}{2} x f(x)$$

$$- \frac{1}{2}(x^2 - 2x) f'(x) - \frac{1}{2}(2x - 2) f(x) - \frac{1}{2}(x - 2) f(x)$$

$$= \frac{1}{2} \left[x^2 f'(x) + x f(x) - 2x f'(x) - 2f(x) + x f(x) + x f(x) - x^2 f'(x) \right.$$

$$\left. + 2x f'(x) - 2x f(x) + 2f(x) - x f(x) + 2f(x) \right] = f(x)$$

Thus, $y_p(x)$ satisfies the differential equation. To see that the boundary conditions are satisfied we first compute

$$y_p(0) = \frac{1}{2}(0 - 2) \int_0^0 t f(t) \, dt + \frac{1}{2}(0) \int_0^1 (t - 2) f(t) \, dt = 0.$$

Next we use $y_p'(x)$ found at the beginning of this part of the solution to compute

$$y_p(1) + y_p'(1) = \frac{1}{2}(1-2)\int_0^1 tf(t)\,dt + \frac{1}{2}(1)\int_1^1 (t-2)f(t)\,dt + \frac{1}{2}(1-2)f(x)$$

$$+ \frac{1}{2}\int_0^1 tf(t)\,dt + \frac{1}{2}(1)[-(1-2)f(x)] + \frac{1}{2}(1)\int_1^1 (1-2)f(t)\,dt$$

$$= \frac{1}{2}(-1)\int_0^1 tf(t)\,dt + \frac{1}{2}(-1)f(1) + \frac{1}{2}\int_0^1 tf(t)\,dt + \frac{1}{2}f(1) = 0.$$

39. The boundary-value problem is $y'' + y = 1$, $\quad y(0) = 0$, $y(1) = 0$. The solution of the associated homogeneous equation is $y = c_1 \cos x + c_2 \sin x$. Since $y(0) = c_1 \cos 0 + c_2 \sin 0 = c_1 = 0$, we take $y_1(x) = \sin x$. To satisfy $y(1) = 0$ we note that $y(1) = c_1 \cos 1 + c_2 \sin 1 = 0$ which implies that $c_1 = -c_2 \sin 1 / \cos 1$ so

$$y(x) = -c_2 \frac{\sin 1}{\cos 1} \cos x + c_2 \sin x = -\frac{c_2}{\cos 1}(\sin x \cos 1 - \cos x \sin 1).$$

Taking $c_2 = -\cos 1$, we have

$$y_2(x) = \sin x \cos 1 - \cos x \sin 1 = \sin(x-1).$$

The Wronskian of y_1 and y_2 is

$$W(y_1, y_2) = \begin{vmatrix} \sin x & \sin(x-1) \\ \cos x & \cos(x-1) \end{vmatrix} = \sin x \cos(x-1) - \cos x \sin(x-1) = \sin[x-(x-1)] = \sin 1,$$

so

$$G(x, t) = \begin{cases} \dfrac{\sin t \sin(x-1)}{\sin 1}, & 0 \le t \le x \\[2mm] \dfrac{\sin x \sin(t-1)}{\sin 1}, & x \le t \le 1. \end{cases}$$

Therefore, taking $f(t) = 1$,

$$y_p(x) = \int_0^1 G(x, t)\,dt = \frac{\sin(x-1)}{\sin 1}\int_0^x \sin t\,dt + \frac{\sin x}{\sin 1}\int_x^1 \sin(t-1)\,dt$$

$$= \frac{\sin(x-1)}{\sin 1}(-\cos x + 1) + \frac{\sin x}{\sin 1}[-1 + \cos(x-1)] = \frac{\sin(x-1)}{\sin 1} - \frac{\sin x}{\sin 1} + 1.$$

42. The boundary-value problem is $y'' - y' = e^{2x}$, $y(0) = 0$, $y(1) = 0$. The solution of the associated homogeneous equation is $y = c_1 + c_2 e^x$. Since $y(0) = c_1 + c_2 = 0$, we see that $c_1 = -c_2$ and we take $y_1(x) = 1 - e^x$. To satisfy $y(1) = 0$ we note that $y(1) = c_1 + c_2 e = 0$ which implies that $c_1 = -c_2 e$ so

$$y(x) = -c_2 e + c_2 e^x = -c_2 e(1 - e^{x-1}).$$

Taking $c_2 = -1/e$, we have $y_2(x) = 1 - e^{x-1}$. The Wronskian of y_1 and y_2 is

$$W(y_1, y_2) = \begin{vmatrix} 1 - e^x & 1 - e^{x-1} \\ -e^x & -e^{x-1} \end{vmatrix} = -e^{x-1} + e^{2x-1} + e^x - e^{2x-1} = e^x - e^{x-1} = e^x(1 - e^{-1}),$$

so

$$G(x, t) = \begin{cases} \dfrac{(1 - e^t)(1 - e^{x-1})}{e^t(1 - e^{-1})}, & 0 \le t \le x \\[2mm] \dfrac{(1 - e^x)(1 - e^{t-1})}{e^t(1 - e^{-1})}, & x \le t \le 1. \end{cases}$$

Therefore, taking $f(t) = e^{2t}$,

$$y_p(x) = \int_0^1 G(x, t)\, e^{2t}\, dt = \frac{1 - e^{x-1}}{1 - e^{-1}} \int_0^x (e^t - e^{2t})\, dt + \frac{1 - e^x}{1 - e^{-1}} \int_x^1 (e^t - e^{2t-1})\, dt$$

$$= \frac{1 - e^{x-1}}{1 - e^{-1}} \left(e^x - \frac{1}{2} e^{2x} - \frac{1}{2} \right) + \frac{1 - e^x}{1 - e^{-1}} \left(\frac{1}{2} e - e^x + \frac{1}{2} e^{2x-1} \right)$$

$$= \frac{1}{2} e^{2x} - \frac{1}{2} e^x = \frac{1}{2} e^{x+1} + \frac{1}{2} e.$$

4.9   Solving Systems of Linear DEs by Elimination

The terminology and concepts listed below provide an outline of the main ideas encountered in this section. These can be useful when preparing for a quiz or test.

Terminology and Concepts

- system of linear DEs

- systematic elimination to find the solution of a system of linear DEs

The basic skills listed below summarize the more mechanical types of problems encountered in the exercise set for this section.

Basic Skills

- use systematic elimnation to solve a system of linear DEs

3. From $Dx = -y + t$ and $Dy = x - t$ we obtain $y = t - Dx$, $Dy = 1 - D^2x$, and $(D^2 + 1)x = 1 + t$. The solution is

$$x = c_1 \cos t + c_2 \sin t + 1 + t$$
$$y = c_1 \sin t - c_2 \cos t + t - 1.$$

6. From $(D + 1)x + (D - 1)y = 2$ and $3x + (D + 2)y = -1$ we obtain $x = -\frac{1}{3} - \frac{1}{3}(D + 2)y$, $Dx = -\frac{1}{3}(D^2 + 2D)y$, and $(D^2 + 5)y = -7$. The solution is

$$y = c_1 \cos \sqrt{5}\,t + c_2 \sin \sqrt{5}\,t - \frac{7}{5}$$

$$x = \left(-\frac{2}{3}c_1 - \frac{\sqrt{5}}{3}c_2\right) \cos \sqrt{5}\,t + \left(\frac{\sqrt{5}}{3}c_1 - \frac{2}{3}c_2\right) \sin \sqrt{5}\,t + \frac{3}{5}.$$

9. From $Dx + D^2y = e^{3t}$ and $(D + 1)x + (D - 1)y = 4e^{3t}$ we obtain $D(D^2 + 1)x = 34e^{3t}$ and $D(D^2 + 1)y = -8e^{3t}$. The solution is

$$y = c_1 + c_2 \sin t + c_3 \cos t - \frac{4}{15}e^{3t}$$

$$x = c_4 + c_5 \sin t + c_6 \cos t + \frac{17}{15}e^{3t}.$$

Substituting into $(D + 1)x + (D - 1)y = 4e^{3t}$ gives

$$(c_4 - c_1) + (c_5 - c_6 - c_3 - c_2)\sin t + (c_6 + c_5 + c_2 - c_3)\cos t = 0$$

so that $c_4 = c_1$, $c_5 = c_3$, $c_6 = -c_2$, and

$$x = c_1 - c_2 \cos t + c_3 \sin t + \frac{17}{15}e^{3t}.$$

12. From $(2D^2 - D - 1)x - (2D + 1)y = 1$ and $(D - 1)x + Dy = -1$; $(2D + 1)(D - 1)(D + 1)x = -1$ and $(2D + 1)(D + 1)y = -2$. The solution is

$$x = c_1 e^{-t/2} + c_2 e^{-t} + c_3 e^t + 1$$
$$y = c_4 e^{-t/2} + c_5 e^{-t} - 2.$$

Substituting into $(D-1)x + Dy = -1$ gives

$$\left(-\frac{3}{2}c_1 - \frac{1}{2}c_4\right)e^{-t/2} + (-2c_2 - c_5)e^{-t} = 0$$

so that $c_4 = -3c_1$, $c_5 = -2c_2$, and

$$y = -3c_1 e^{-t/2} - 2c_2 e^{-t} - 2.$$

15. Multiplying the first equation by $D+1$ and the second equation by D^2+1 and subtracting we obtain $(D^4 - D^2)x = 1$. Then

$$x = c_1 + c_2 t + c_3 e^t + c_4 e^{-t} - \frac{1}{2}t^2.$$

Multiplying the first equation by $D+1$ and subtracting we obtain $D^2(D+1)y = 1$. Then

$$y = c_5 + c_6 t + c_7 e^{-t} - \frac{1}{2}t^2.$$

Substituting into $(D-1)x + (D^2+1)y = 1$ gives

$$(-c_1 + c_2 + c_5 - 1) + (-2c_4 + 2c_7)e^{-t} + (-1 - c_2 + c_6)t = 1$$

so that $c_5 = c_1 - c_2 + 2$, $c_6 = c_2 + 1$, and $c_7 = c_4$. The solution of the system is

$$x = c_1 + c_2 t + c_3 e^t + c_4 e^{-t} - \frac{1}{2}t^2$$

$$y = (c_1 - c_2 + 2) + (c_2 + 1)t + c_4 e^{-t} - \frac{1}{2}t^2.$$

18. From $Dx + z = e^t$, $(D-1)x + Dy + Dz = 0$, and $x + 2y + Dz = e^t$ we obtain $z = -Dx + e^t$, $Dz = -D^2 x + e^t$, and the system $(-D^2 + D - 1)x + Dy = -e^t$ and $(-D^2 + 1)x + 2y = 0$. Then $y = \frac{1}{2}(D^2 - 1)x$, $Dy = \frac{1}{2}D(D^2 - 1)x$, and $(D-2)(D^2+1)x = -2e^t$ so that the solution is

$$x = c_1 e^{2t} + c_2 \cos t + c_3 \sin t + e^t$$

$$y = \frac{3}{2}c_1 e^{2t} - c_2 \cos t - c_3 \sin t$$

$$z = -2c_1 e^{2t} - c_3 \cos t + c_2 \sin t.$$

21. From $(D+5)x + y = 0$ and $4x - (D+1)y = 0$ we obtain $y = -(D+5)x$ so that $Dy = -(D^2+5D)x$. Then $4x + (D^2+5D)x + (D+5)x = 0$ and $(D+3)^2 x = 0$. Thus

$$x = c_1 e^{-3t} + c_2 t e^{-3t}$$

$$y = -(2c_1 + c_2)e^{-3t} - 2c_2 t e^{-3t}.$$

Using $x(1) = 0$ and $y(1) = 1$ we obtain

$$c_1 e^{-3} + c_2 e^{-3} = 0$$

$$-(2c_1 + c_2)e^{-3} - 2c_2 e^{-3} = 1$$

or

$$c_1 + c_2 = 0$$

$$2c_1 + 3c_2 = -e^3.$$

Thus $c_1 = e^3$ and $c_2 = -e^3$. The solution of the initial value problem is

$$x = e^{-3t+3} - te^{-3t+3}$$

$$y = -e^{-3t+3} + 2te^{-3t+3}.$$

Mathematical Models

24. From Newton's second law in the x-direction we have

$$m\frac{d^2 x}{dt^2} = -k\cos\theta = -k\frac{1}{v}\frac{dx}{dt} = -|c|\frac{dx}{dt}.$$

In the y-direction we have

$$m\frac{d^2 y}{dt^2} = -mg - k\sin\theta = -mg - k\frac{1}{v}\frac{dy}{dt} = -mg - |c|\frac{dy}{dt}.$$

From $mD^2 x + |c|Dx = 0$ we have $D(mD + |c|)x = 0$. Then $(mD + |c|)x = c_1$ or $(D + |c|/m)x = c_2$. This is a linear first-order differential equation. An integrating factor is $e^{\int |c|dt/m} = e^{|c|t/m}$ so that

$$\frac{d}{dt}[e^{|c|t/m}x] = c_2 e^{|c|t/m}$$

and $e^{|c|t/m}x = (c_2 m/|c|)e^{|c|t/m} + c_3$. The general solution of this equation is

$$x(t) = c_4 + c_3 e^{-|c|t/m}.$$

From $(mD^2+|c|D)y = -mg$ we have $D(mD+|c|)y = -mg$ so that $(mD+|c|)y = -mgt+c_1$ or $(D + |c|/m)y = -gt + c_2$. This is a linear first-order differential equation with integrating factor $e^{\int |c|dt/m} = e^{|c|t/m}$. Thus

$$\frac{d}{dt}[e^{|c|t/m}y] = (-gt + c_2)e^{|c|t/m}$$

$$e^{|c|t/m}y = -\frac{mg}{|c|}\,te^{|c|t/m} + \frac{m^2g}{c^2}\,e^{|c|t/m} + c_3e^{|c|t/m} + c_4$$

and

$$y(t) = -\frac{mg}{|c|}\,t + \frac{m^2g}{c^2} + c_3 + c_4e^{-|c|t/m}.$$

4.10 | Nonlinear Differential Equations

The terminology and concepts listed below provide an outline of the main ideas encountered in this section. These can be useful when preparing for a quiz or test.

Terminology and Concepts

- superposition of solutions does *not* apply to nonlinear DEs

- nonlinear DEs can possess singular solutions

- using a substitution, express a nonlinear second-order DE, $F(x, y', y'') = 0$, as a first-order DE

- using a substitution, express a nonlinear second-order DE, $F(y, y', y'') = 0$, as a first-order DE

- Taylor series solution of an IVP

The basic skills listed below summarize the more mechanical types of problems encountered in the exercise set for this section.

Basic Skills

- solve a nonlinear second-order DE using reduction of order when appropriate

- use a substitution to find a Taylor series solution of a nonlinear second-order IVP

3. Let $u = y'$ so that $u' = y''$. The equation becomes $u' = -u^2 - 1$ which is separable. Thus

$$\frac{du}{u^2 + 1} = -dx \quad \Longrightarrow \quad \tan^{-1} u = -x + c_1 \quad \Longrightarrow \quad y' = \tan(c_1 - x)$$

$$\Longrightarrow \quad y = \ln|\cos(c_1 - x)| + c_2.$$

6. Let $u = y'$ so that $y'' = u\,du/dy$. The equation becomes $(y+1)u\,du/dy = u^2$. Separating variables we obtain

$$\frac{du}{u} = \frac{dy}{y+1} \quad\Longrightarrow\quad \ln|u| = \ln|y+1| + \ln c_1 \quad\Longrightarrow\quad u = c_1(y+1)$$

$$\Longrightarrow\quad \frac{dy}{dx} = c_1(y+1) \quad\Longrightarrow\quad \frac{dy}{y+1} = c_1\,dx$$

$$\Longrightarrow\quad \ln|y+1| = c_1 x + c_2 \quad\Longrightarrow\quad y+1 = c_3 e^{c_1 x}.$$

9. Letting $u = y'$ we have

$$y'' = \frac{du}{dx} = \frac{du}{dy}\frac{dy}{dx} = u\frac{du}{dy} \qquad\text{so}\qquad 2y'y'' = 1 \qquad\text{becomes}\qquad 2u^2\frac{du}{dy} = 1.$$

Then, separating variables, integrating, and simplifying, we have

$$2u^2\,du = dy$$

$$\frac{2}{3}u^3 = y + c_1$$

$$u = \left(\frac{3}{2}y + c_2\right)^{1/3} = \frac{dy}{dx}$$

$$\left(\frac{3}{2}y + c_2\right)^{-1/3}dy = dx$$

$$\left(\frac{3}{2}y + c_2\right)^{2/3} = x + c_3$$

$$\frac{3}{2}y + c_2 = \left(x + c_3\right)^{3/2}.$$

Now

$$y(0) = 2 \quad\text{implies}\quad 3 + c_2 = c_3^{3/2} \qquad\text{and}\qquad y'(0) = 1 \quad\text{implies}\quad c_3 = 1.$$

Thus $c_2 = -2$ and $\frac{3}{2}y - 2 = (x+1)^{3/2}$. The solution of the initial-value problem is

$$y = \frac{2}{3}(x+1)^{3/2} + \frac{4}{3}.$$

12. Let $u = y'$ so that $u' = y''$. The equation becomes $(u')^2 + u^2 = 1$ which results in $u' = \pm\sqrt{1 - u^2}$. To solve $u' = \sqrt{1 - u^2}$ we separate variables:

$$\frac{du}{\sqrt{1 - u^2}} = dx \quad\Longrightarrow\quad \sin^{-1}u = x + c_1 \quad\Longrightarrow\quad u = \sin(x + c_1)$$

$$\Longrightarrow\quad y' = \sin(x + c_1).$$

When $x = \pi/2$, $y' = \sqrt{3}/2$, so $\sqrt{3}/2 = \sin(\pi/2 + c_1)$ and $c_1 = -\pi/6$. Thus

$$y' = \sin\left(x - \frac{\pi}{6}\right) \implies y = -\cos\left(x - \frac{\pi}{6}\right) + c_2.$$

When $x = \pi/2$, $y = 1/2$, so $1/2 = -\cos(\pi/2 - \pi/6) + c_2 = -1/2 + c_2$ and $c_2 = 1$. The solution of the initial-value problem is $y = 1 - \cos(x - \pi/6)$.

To solve $u' = -\sqrt{1 - u^2}$ we separate variables:

$$\frac{du}{\sqrt{1 - u^2}} = -dx \implies \cos^{-1} u = x + c_1$$

$$\implies u = \cos(x + c_1) \implies y' = \cos(x + c_1).$$

When $x = \pi/2$, $y' = \sqrt{3}/2$, so $\sqrt{3}/2 = \cos(\pi/2 + c_1)$ and $c_1 = -\pi/3$. Thus

$$y' = \cos\left(x - \frac{\pi}{3}\right) \implies y = \sin\left(x - \frac{\pi}{3}\right) + c_2.$$

When $x = \pi/2$, $y = 1/2$, so $1/2 = \sin(\pi/2 - \pi/3) + c_2 = 1/2 + c_2$ and $c_2 = 0$. The solution of the initial-value problem is $y = \sin(x - \pi/3)$.

In Problems 15 and 18 the thinner curve is obtained using a numerical solver, while the thicker curve is the graph of the Taylor polynomial.

15. We look for a solution of the form

$$y(x) = y(0) + y'(0)x + \frac{1}{2!}y''(0)x^2 + \frac{1}{3!}y'''(0)x^3 +$$

$$\frac{1}{4!}y^{(4)}(0)x^4 + \frac{1}{5!}y^{(5)}(0)x^5.$$

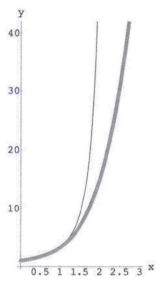

From $y''(x) = x + y^2$ we compute

$$y'''(x) = 1 + 2yy'$$

$$y^{(4)}(x) = 2yy'' + 2(y')^2$$

$$y^{(5)}(x) = 2yy''' + 6y'y''.$$

Using $y(0) = 1$ and $y'(0) = 1$ we find

$$y''(0) = 1, \quad y'''(0) = 3, \quad y^{(4)}(0) = 4, \quad y^{(5)}(0) = 12.$$

An approximate solution is

$$y(x) = 1 + x + \frac{1}{2}x^2 + \frac{1}{2}x^3 + \frac{1}{6}x^4 + \frac{1}{10}x^5.$$

18. We look for a solution of the form

$$y(x) = y(0) + y'(0)x + \frac{1}{2!}y''(0)x^2 + \frac{1}{3!}y'''(0)x^3 +$$

$$\frac{1}{4!}y^{(4)}(0)x^4 + \frac{1}{5!}y^{(5)}(0)x^5 + \frac{1}{6!}y^{(6)}(0)x^6.$$

From $y''(x) = e^y$ we compute

$$y'''(x) = e^y y'$$
$$y^{(4)}(x) = e^y(y')^2 + e^y y''$$
$$y^{(5)}(x) = e^y(y')^3 + 3e^y y' y'' + e^y y'''$$
$$y^{(6)}(x) = e^y(y')^4 + 6e^y(y')^2 y'' + 3e^y(y'')^2 + 4e^y y' y''' + e^y y^{(4)}.$$

Using $y(0) = 0$ and $y'(0) = -1$ we find

$$y''(0) = 1, \quad y'''(0) = -1, \quad y^{(4)}(0) = 2, \quad y^{(5)}(0) = -5, \quad y^{(6)}(0) = 16.$$

An approximate solution is

$$y(x) = -x + \frac{1}{2}x^2 - \frac{1}{6}x^3 + \frac{1}{12}x^4 + \frac{1}{24}x^5 + \frac{1}{45}x^6.$$

4.R Chapter 4 in Review

3. False; it is not true unless the differential equation is homogeneous. For example, $y_1 = x$ is a solution of $y'' + y = x$, but $y_2 = 5x$ is not.

6. The roots are $m_1 = m_2 = m_3 = 0$ and $m_4 = 1$. To see this, note that the general solution of a homogeneous linear fourth-order differential equation with auxiliary equation $m^3(m-1) = 0$ is $y = c_1 + c_2x + c_3x^2 + c_4e^x$.

9. By the superposition principle for nonhomogeneous equations a particular solution is

$$y_p = y_{p_1} + y_{p_2} = x + x^2 - 2 = x^2 + x - 2.$$

12. (a) Since $f_2(x) = 2\ln x = 2f_1(x)$, the set of functions is linearly dependent.

(b) Since x^{n+1} is not a constant multiple of x^n, the set of functions is linearly independent.

(c) Since $x + 1$ is not a constant multiple of x, the set of functions is linearly independent.

(d) Since $f_1(x) = \cos x \cos(\pi/2) - \sin x \sin(\pi/2) = -\sin x = -f_2(x)$, the set of functions is linearly dependent.

(e) Since $f_1(x) = 0 \cdot f_2(x)$, the set of functions is linearly dependent.

(f) Since $2x$ is not a constant multiple of 2, the set of functions is linearly independent.

(g) Since $3(x^2) + 2(1 - x^2) - (2 + x^2) = 0$, the set of functions is linearly dependent.

(h) Since $xe^{x+1} + 0(4x - 5)e^x - exe^x = 0$, the set of functions is linearly dependent.

15. From $m^2 - 2m - 2 = 0$ we obtain $m = 1 \pm \sqrt{3}$ so that

$$y = c_1 e^{(1+\sqrt{3})x} + c_2 e^{(1-\sqrt{3})x}.$$

18. From $2m^3 + 9m^2 + 12m + 5 = 0$ we obtain $m = -1$, $m = -1$, and $m = -5/2$ so that

$$y = c_1 e^{-5x/2} + c_2 e^{-x} + c_3 x e^{-x}.$$

21. Applying D^4 to the differential equation we obtain $D^4(D^2 - 3D + 5) = 0$. Then

$$y = \underbrace{e^{3x/2}\left(c_1 \cos \frac{\sqrt{11}}{2}x + c_2 \sin \frac{\sqrt{11}}{2}x\right)}_{y_c} + c_3 + c_4 x + c_5 x^2 + c_6 x^3$$

and $y_p = A + Bx + Cx^2 + Dx^3$. Substituting y_p into the differential equation yields

$$(5A - 3B + 2C) + (5B - 6C + 6D)x + (5C - 9D)x^2 + 5Dx^3 = -2x + 4x^3.$$

Equating coefficients gives $A = -222/625$, $B = 46/125$, $C = 36/25$, and $D = 4/5$. The general solution is

$$y = e^{3x/2}\left(c_1 \cos \frac{\sqrt{11}}{2}x + c_2 \sin \frac{\sqrt{11}}{2}x\right) - \frac{222}{625} + \frac{46}{125}x + \frac{36}{25}x^2 + \frac{4}{5}x^3.$$

24. Applying D to the differential equation we obtain $D(D^3 - D^2) = D^3(D - 1) = 0$. Then

$$y = \underbrace{c_1 + c_2 x + c_3 e^x}_{y_c} + c_4 x^2$$

and $y_p = Ax^2$. Substituting y_p into the differential equation yields $-2A = 6$. Equating coefficients gives $A = -3$. The general solution is

$$y = c_1 + c_2 x + c_3 e^x - 3x^2.$$

27. The auxiliary equation is $6m^2 - m - 1 = 0$ so that

$$y = c_1 x^{1/2} + c_2 x^{-1/3}.$$

30. The auxiliary equation is $m^2 - 2m + 1 = (m-1)^2 = 0$ and a particular solution is $y_p = \frac{1}{4}x^3$ so that

$$y = c_1 x + c_2 x \ln x + \frac{1}{4}x^3.$$

33. (a) The auxiliary equation is $m^4 - 2m^2 + 1 = (m^2 - 1)^2 = 0$, so the general solution of the differential equation is

$$y = c_1 \sinh x + c_2 \cosh x + c_3 x \sinh x + c_4 x \cosh x.$$

(b) Since both $\sinh x$ and $x \sinh x$ are solutions of the associated homogeneous differential equation, a particular solution of $y^{(4)} - 2y'' + y = \sinh x$ has the form

$$y_p = Ax^2 \sinh x + Bx^2 \cosh x.$$

36. The auxiliary equation is $m^2 + 2m + 1 = (m+1)^2 = 0$, so that $y = c_1 e^{-x} + c_2 x e^{-x}$. Setting $y(-1) = 0$ and $y'(0) = 0$ we get $c_1 e - c_2 e = 0$ and $-c_1 + c_2 = 0$. Thus $c_1 = c_2$ and $y = c_1(e^{-x} + xe^{-x})$ is a solution of the boundary-value problem for any real number c_1.

39. Let $u = y'$ so that $u' = y''$. The equation becomes $u\, du/dx = 4x$. Separating variables we obtain

$$u\, du = 4x\, dx \quad \Longrightarrow \quad \frac{1}{2}u^2 = 2x^2 + c_1 \quad \Longrightarrow \quad u^2 = 4x^2 + c_2.$$

When $x = 1$, $y' = u = 2$, so $4 = 4 + c_2$ and $c_2 = 0$. Then

$$u^2 = 4x^2 \quad \Longrightarrow \quad \frac{dy}{dx} = 2x \quad \text{or} \quad \frac{dy}{dx} = -2x$$

$$\Longrightarrow \quad y = x^2 + c_3 \quad \text{or} \quad y = -x^2 + c_4.$$

When $x = 1$, $y = 5$, so $5 = 1 + c_3$ and $5 = -1 + c_4$. Thus $c_3 = 4$ and $c_4 = 6$. We have $y = x^2 + 4$ and $y = -x^2 + 6$. Note however that when $y = -x^2 + 6$, $y' = -2x$ and $y'(1) = -2 \neq 2$. Thus, the solution of the initial-value problem is $y = x^2 + 4$.

42. Consider $xy'' + y' = 0$ and look for a solution of the form $y = x^m$. Substituting into the differential equation we have

$$xy'' + y' = m(m-1)x^{m-1} + mx^{m-1} = m^2 x^{m-1}.$$

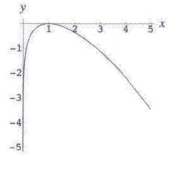

Thus, the general solution of $xy'' + y' = 0$ is $y_c = c_1 + c_2 \ln x$. To find a particular solution of $xy'' + y' = -\sqrt{x}$ we use variation of parameters. The Wronskian is $W = 1/x$. Identifying $f(x) = -x^{-1/2}$ we obtain

$$u_1' = \frac{x^{-1/2} \ln x}{1/x} = \sqrt{x} \ln x \quad \text{and} \quad u_2' = \frac{-x^{-1/2}}{1/x} = -\sqrt{x},$$

so that

$$u_1 = x^{3/2}\left(\frac{2}{3}\ln x - \frac{4}{9}\right) \quad \text{and} \quad u_2 = -\frac{2}{3}x^{3/2}.$$

Then

$$y_p = x^{3/2}\left(\frac{2}{3}\ln x - \frac{4}{9}\right) - \frac{2}{3}x^{3/2}\ln x = -\frac{4}{9}x^{3/2}$$

and the general solution of the differential equation is

$$y = c_1 + c_2 \ln x - \frac{4}{9}x^{3/2}.$$

The initial conditions are $y(1) = 0$ and $y'(1) = 0$. These imply that $c_1 = \frac{4}{9}$ and $c_2 = \frac{2}{3}$. The solution of the initial-value problem is

$$y = \frac{4}{9} + \frac{2}{3}\ln x - \frac{4}{9}x^{3/2}.$$

The graph is shown above.

45. From $(D-2)x - y = -e^t$ and $-3x + (D-4)y = -7e^t$ we obtain $(D-1)(D-5)x = -4e^t$ so that

$$x = c_1 e^t + c_2 e^{5t} + te^t.$$

Then

$$y = (D-2)x + e^t = -c_1 e^t + 3c_2 e^{5t} - te^t + 2e^t.$$

5 Modeling with Higher-Order Differential Equations

5.1 | Linear Models: Initial-Value Problems

The terminology and concepts listed below provide an outline of the main ideas encountered in this section. These can be useful when preparing for a quiz or test.

Terminology and Concepts

- Hooke's law

- spring constant (k)

- Newton's second law of motion

- mass (m) versus weight ($W = mg$)

- equilibrium position

- DE of free undamped motion: $m\,d^2x/dt^2 + kx = 0$ or $d^2x/dt^2 + \omega^2 x = 0$, where $\omega^2 = k/m$

- equation of free undamped (simple harmonic) motion: $x(t) = c_1 \cos \omega t + c_2 \sin \omega t$

- period of motion: $T = 2\pi/\omega$

- cycle

- frequency of motion: $f = \omega/2\pi$

- amplitude of free vibrations (A)

- phase angle (ϕ)

- damping constant (β)

- DE of free damped motion: $m\,d^2x/dt^2 + \beta\,dx/dt + kx = 0$ or $d^2x/dt^2 + 2\lambda\,dx/dt + \omega^2 x = 0$ where $2\lambda = \beta/m$ and $\omega^2 = k/m$

- equation of motion of an overdamped system: $x(t) = e^{-\lambda t}(c_1 e^{\sqrt{\lambda^2 - \omega^2}\,t} + c_2 e^{-\sqrt{\lambda^2 - \omega^2}\,t})$

- equation of motion of a critically damped system: $x(t) = e^{-\lambda t}(c_1 + c_2 t)$

- equation of motion of an underdamped system:

$$x(t) = e^{-\lambda t}\left(c_1 \cos \sqrt{\omega^2 - \lambda^2}\, t + c_2 \sin \sqrt{\omega^2 - \lambda^2}\, t\right)$$

- damped amplitude: $Ae^{-\lambda t}$

- quasi period: $2\pi/\sqrt{\omega^2 - \lambda^2}$

- DE of driven (forced) motion with damping: $m\, d^2x/dt^2 + \beta\, dx/dt + kx = f(t)$

- transient term (solution)

- steady-state term (solution)

- DE of driven motion without damping: $m\, d^2x/dt^2 + kx = f(t)$

- pure resonance

- aging spring

- Airy's differential equation

- *LRC* series electrical circuit

- free electrical vibrations

- reactance

- impedance

The basic skills listed below summarize the more mechanical types of problems encountered in the exercise set for this section.

Basic Skills

- set up and solve DEs modeling free undamped motion of a spring/mass system

- set up and solve DEs modeling free damped motion of a spring/mass system

- set up and solve DEs modeling driven (forced) motion of a spring/mass system

- set up and solve DEs modeling an *LRC* series circuit

Phase Angle   As pointed out in the text, be careful in computing the phase angle ϕ in a solution of the form $x(t) = A\sin(\omega t + \phi)$. The problem with using the inverse tangent $\tan^{-1} X$ is that all calculators are programmed for this particular inverse trigonometric function to give angular values that are either in the first or fourth quadrants. In other words, $-\pi/2 < \tan^{-1} X < \pi/2$. For example, because of periodicity there are infinitely many angles for which $\tan \phi = 1$ and for which $\tan \phi = -1/\sqrt{3}$. But there is only *one* answer when we use the notation $\tan^{-1} 1$ and $\tan^{-1}(-1/\sqrt{3})$. The exact values are:

$$\tan^{-1} 1 = \frac{\pi}{4} \quad \text{(first quadrant) because } \tan \frac{\pi}{4} = 1,$$

$$\tan^{-1}\left(-\frac{1}{\sqrt{3}}\right) = -\frac{\pi}{6} \quad \text{(fourth quadrant) because} \quad \tan\left(-\frac{\pi}{6}\right) = -\frac{1}{\sqrt{3}}.$$

Of course, if you use a calculator you will either get

$$\tan^{-1} 1 = 45° \quad \text{or} \quad \tan^{-1} 1 = 0.78539816\ldots$$

depending on whether your calculator is set in degree or radian mode. The decimal number $0.78539816\ldots$ is the number π divided by 4. Now in order to compute ϕ in $x(t) = A\sin(\omega t + \phi)$ we must take into consideration all three components given in (7) of Section 5.1 in the text. Since $A > 0$, the signs of $\sin\phi = c_1/A$ and $\cos\phi = c_2/A$ are determined from the algebraic signs of c_1 and c_2 and these signs, in turn, indicate the quadrant in which ϕ lies. We summarize the four possibilities:

$$\phi = \tan^{-1}\left(\frac{c_1}{c_2}\right) \quad \text{if} \quad \overbrace{c_1 > 0,\; c_2 > 0}^{\phi \text{ in 1st quadrant}} \quad \text{or if} \quad \overbrace{c_1 < 0,\; c_2 > 0}^{\phi \text{ in 4th quadrant}}$$

$$\phi = \pi + \tan^{-1}\left(\frac{c_1}{c_2}\right) \quad \text{if} \quad \overbrace{c_1 > 0,\; c_2 < 0}^{\phi \text{ in 2nd quadrant}} \quad \text{or if} \quad \overbrace{c_1 < 0,\; c_2 < 0}^{\phi \text{ in 3rd quadrant}}.$$

Damping In Examples 4 and 5 in the text, the statements

"...a damping force numerically equal to 2 times the instantaneous velocity..."

and

"...surrounding medium offers a resistance numerically equal to the instantaneous velocity..."

are to be interpreted: In Example 4, the damping constant is $\beta = 2$ in equation (10) and in Example 5, the damping constant is $\beta = 1$ in equation (10).

5.1.1 Spring/Mass Systems: Free Undamped Motion

3. From $\frac{3}{4}x'' + 72x = 0$, $x(0) = -1/4$, and $x'(0) = 0$ we obtain $x = -\frac{1}{4}\cos 4\sqrt{6}\,t$.

6. From $50x'' + 200x = 0$, $x(0) = 0$, and $x'(0) = -10$ we obtain $x = -5\sin 2t$ and $x' = -10\cos 2t$.

9. (a) Since the weight is 8 pounds the mass is 8/32=1/4 slug. The spring constant is $k = 1$ so $\omega^2 = 1/\frac{1}{4} = 4$ and the equation of motion is

$$\frac{1}{4}x'' + x = 0 \quad \text{or} \quad x'' + 4x = 0.$$

From the initial conditions we have $x(0) = 6$ in $= \frac{1}{2}$ ft and $x'(0) = \frac{3}{2}$ ft/sec. By (3) in the text the form of the equation of motion is

$$x(t) = c_1 \cos 2t + c_2 \sin 2t.$$

The initial conditions imply that $c_1 = \frac{1}{2}$ and $c_2 = \frac{3}{4}$. Thus the equation of motion of the system is

$$x(t) = \frac{1}{2} \cos 2t + \frac{3}{4} \sin 2t.$$

(b) To write the solution in the form $x(t) = A \sin(\omega t + \phi)$ we note that

$$A = \sqrt{\left(\frac{1}{2}\right)^2 + \left(\frac{3}{4}\right)^2} = \frac{1}{4}\sqrt{4 + 9} = \frac{\sqrt{13}}{4},$$

and, initially at least,

$$\phi = \tan^{-1}\frac{c_1}{c_2} = \tan^{-1}\frac{1/2}{3/4} = \tan^{-1}\frac{2}{3} \approx 0.5880 \text{ rad.}$$

Since $\sin\phi > 0$ and $\cos\phi > 0$, ϕ is in the first quadrant and

$$x(t) = \frac{\sqrt{13}}{4} \sin(2t + 0.5880).$$

(c) To write the system in the form $x(t) = A\cos(\omega t - \phi)$ we note that $A = \sqrt{13}/4$ as in part **(b)** and

$$\phi = \tan^{-1}\frac{3/4}{1/2} = \frac{3}{2} \approx 0.9828.$$

Since $\cos\phi > 0$ and $\sin\phi > 0$, ϕ is in the first quadrant and

$$x(t) = \frac{\sqrt{13}}{4} \cos(2t - 0.9828).$$

12. From $x'' + 9x = 0$, $x(0) = -1$, and $x'(0) = -\sqrt{3}$ we obtain

$$x = -\cos 3t - \frac{\sqrt{3}}{3} \sin 3t = \frac{2}{\sqrt{3}} \sin\left(3t + \frac{4\pi}{3}\right)$$

and $x' = 2\sqrt{3}\cos(3t + 4\pi/3)$. If $x' = 3$ then $t = -7\pi/18 + 2n\pi/3$ and $t = -\pi/2 + 2n\pi/3$ for $n = 1, 2, 3, \ldots$.

15. For large values of t the differential equation is approximated by $x'' = 0$. The solution of this equation is the linear function $x = c_1 t + c_2$. Thus, for large time, the restoring force will have decayed to the point where the spring is incapable of returning the mass, and the spring will simply keep on stretching.

5.1.2 Spring/Mass Systems: Free Damped Motion

18. (a) below   **(b)** from rest

21. From $\frac{1}{8}x'' + x' + 2x = 0$, $x(0) = -1$, and $x'(0) = 8$ we obtain $x = 4te^{-4t} - e^{-4t}$ and $x' = 8e^{-4t} - 16te^{-4t}$. If $x = 0$ then $t = 1/4$ second. If $x' = 0$ then $t = 1/2$ second and the extreme displacement is $x = e^{-2}$ feet.

24. (a) $x = \frac{1}{3}e^{-8t}\left(4e^{6t} - 1\right)$ is not zero for $t \geq 0$; the extreme displacement is $x(0) = 1$ meter.

(b) $x = \frac{1}{3}e^{-8t}\left(5 - 2e^{6t}\right) = 0$ when $t = \frac{1}{6}\ln\frac{5}{2} \approx 0.153$ second; if $x' = \frac{4}{3}e^{-8t}\left(e^{6t} - 10\right) = 0$ then $t = \frac{1}{6}\ln 10 \approx 0.384$ second and the extreme displacement is $x = -0.232$ meter.

27. From $\frac{5}{16}x'' + \beta x' + 5x = 0$ we find that the roots of the auxiliary equation are

$$m = -\frac{8}{5}\beta \pm \frac{4}{5}\sqrt{4\beta^2 - 25}.$$

(a) If $4\beta^2 - 25 > 0$ then $\beta > 5/2$.

(b) If $4\beta^2 - 25 = 0$ then $\beta = 5/2$.

(c) If $4\beta^2 - 25 < 0$ then $0 < \beta < 5/2$.

5.1.3 Spring/Mass Systems: Driven Motion

30. (a) If $x'' + 2x' + 5x = 12\cos 2t + 3\sin 2t$, $x(0) = 1$. and $x'(0) = 5$ then

$$x_c = e^{-t}(c_1 \cos 2t + c_2 \sin 2t) \qquad \text{and} \qquad x_p = 3\sin 2t$$

so that the equation of motion is $x = e^{-t}\cos 2t + 3\sin 2t$.

(b)   **(c)**

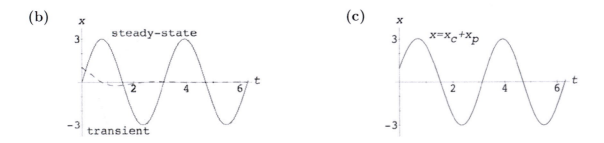

33. From $2x'' + 32x = 68e^{-2t}\cos 4t$, $x(0) = 0$, and $x'(0) = 0$ we obtain $x_c = c_1 \cos 4t + c_2 \sin 4t$ and $x_p = \frac{1}{2}e^{-2t}\cos 4t - 2e^{-2t}\sin 4t$ so that

$$x = -\frac{1}{2}\cos 4t + \frac{9}{4}\sin 4t + \frac{1}{2}e^{-2t}\cos 4t - 2e^{-2t}\sin 4t.$$

36. (a) From $100x'' + 1600x = 1600 \sin 8t$, $x(0) = 0$, and $x'(0) = 0$ we obtain $x_c = c_1 \cos 4t + c_2 \sin 4t$ and $x_p = -\frac{1}{3} \sin 8t$ so that by a trig identity

$$x = \frac{2}{3} \sin 4t - \frac{1}{3} \sin 8t = \frac{2}{3} \sin 4t - \frac{2}{3} \sin 4t \cos 4t.$$

(b) If $x = \frac{1}{3} \sin 4t(2 - 2\cos 4t) = 0$ then $t = n\pi/4$ for $n = 0, 1, 2, \ldots$.

(c) If $x' = \frac{8}{3} \cos 4t - \frac{8}{3} \cos 8t = \frac{8}{3}(1 - \cos 4t)(1 + 2\cos 4t) = 0$ then $t = \pi/3 + n\pi/2$ and $t = \pi/6 + n\pi/2$ for $n = 0, 1, 2, \ldots$ at the extreme values. There are many other values of t for which $x' = 0$.

(d) $x(\pi/6 + n\pi/2) = \sqrt{3}/2$ cm and $x(\pi/3 + n\pi/2) = -\sqrt{3}/2$ cm

(e)

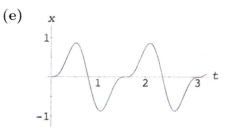

39. (a) From $x'' + \omega^2 x = F_0 \cos \gamma t$, $x(0) = 0$, and $x'(0) = 0$ we obtain $x_c = c_1 \cos \omega t + c_2 \sin \omega t$ and $x_p = (F_0 \cos \gamma t)/(\omega^2 - \gamma^2)$ so that

$$x = -\frac{F_0}{\omega^2 - \gamma^2} \cos \omega t + \frac{F_0}{\omega^2 - \gamma^2} \cos \gamma t.$$

(b) $\displaystyle \lim_{\gamma \to \omega} \frac{F_0}{\omega^2 - \gamma^2}(\cos \gamma t - \cos \omega t) = \lim_{\gamma \to \omega} \frac{-F_0 t \sin \gamma t}{-2\gamma} = \frac{F_0}{2\omega} t \sin \omega t.$

5.1.4 Series Circuit Analogue

45. Solving $\frac{1}{20}q'' + 2q' + 100q = 0$ we obtain $q(t) = e^{-20t}(c_1 \cos 40t + c_2 \sin 40t)$. The initial conditions $q(0) = 5$ and $q'(0) = 0$ imply $c_1 = 5$ and $c_2 = 5/2$. Thus

$$q(t) = e^{-20t}\left(5 \cos 40t + \frac{5}{2} \sin 40t\right) = \sqrt{25 + 25/4}\, e^{-20t} \sin(40t + 1.1071)$$

and $q(0.01) \approx 4.5676$ coulombs. The charge is zero for the first time when $40t + 1.1071 = \pi$ or $t \approx 0.0509$ second.

48. Solving $q'' + 100q' + 2500q = 30$ we obtain $q(t) = c_1 e^{-50t} + c_2 t e^{-50t} + 0.012$. The initial conditions $q(0) = 0$ and $q'(0) = 2$ imply $c_1 = -0.012$ and $c_2 = 1.4$. Thus, using $i(t) = q'(t)$ we get

$$q(t) = -0.012 e^{-50t} + 1.4 t e^{-50t} + 0.012 \quad \text{and} \quad i(t) = 2 e^{-50t} - 70 t e^{-50t}.$$

Solving $i(t) = 0$ we see that the maximum charge occurs when $t = 1/35$ second and $q(1/35) \approx 0.01871$ coulomb.

51. The differential equation is $\frac{1}{2} q'' + 20q' + 1000q = 100 \sin 60t$. To use Example 10 in the text we identify $E_0 = 100$ and $\gamma = 60$. Then

$$X = L\gamma - \frac{1}{c\gamma} = \frac{1}{2}(60) - \frac{1}{0.001(60)} \approx 13.3333,$$

$$Z = \sqrt{X^2 + R^2} = \sqrt{X^2 + 400} \approx 24.0370,$$

and

$$\frac{E_0}{Z} = \frac{100}{Z} \approx 4.1603.$$

From Problem 50, then

$$i_p(t) \approx 4.1603 \sin(60t + \phi)$$

where $\sin \phi = -X/Z$ and $\cos \phi = R/Z$. Thus $\tan \phi = -X/R \approx -0.6667$ and ϕ is a fourth quadrant angle. Now $\phi \approx -0.5880$ and

$$i_p(t) = 4.1603 \sin(60t - 0.5880).$$

54. In Problem 50 it is shown that the amplitude of the steady-state current is E_0/Z, where $Z = \sqrt{X^2 + R^2}$ and $X = L\gamma - 1/C\gamma$. Since E_0 is constant the amplitude will be a maximum when Z is a minimum. Since R is constant, Z will be a minimum when $X = 0$. Solving $L\gamma - 1/C\gamma = 0$ for γ we obtain $\gamma = 1/\sqrt{LC}$. The maximum amplitude will be E_0/R.

57. In an LC-series circuit there is no resistor, so the differential equation is

$$L \frac{d^2 q}{dt^2} + \frac{1}{C} q = E(t).$$

Then $q(t) = c_1 \cos\left(t/\sqrt{LC}\right) + c_2 \sin\left(t/\sqrt{LC}\right) + q_p(t)$ where $q_p(t) = A \sin \gamma t + B \cos \gamma t$. Substituting $q_p(t)$ into the differential equation we find

$$\left(\frac{1}{C} - L\gamma^2\right) A \sin \gamma t + \left(\frac{1}{C} - L\gamma^2\right) B \cos \gamma t = E_0 \cos \gamma t.$$

Equating coefficients we obtain $A = 0$ and $B = E_0 C/(1 - LC\gamma^2)$. Thus, the charge is

$$q(t) = c_1 \cos \frac{1}{\sqrt{LC}} t + c_2 \sin \frac{1}{\sqrt{LC}} t + \frac{E_0 C}{1 - LC\gamma^2} \cos \gamma t.$$

The initial conditions $q(0) = q_0$ and $q'(0) = i_0$ imply $c_1 = q_0 - E_0 C/(1 - LC\gamma^2)$ and $c_2 = i_0\sqrt{LC}$. The current is $i(t) = q'(t)$ or

$$i(t) = -\frac{c_1}{\sqrt{LC}}\sin\frac{1}{\sqrt{LC}}t + \frac{c_2}{\sqrt{LC}}\cos\frac{1}{\sqrt{LC}}t - \frac{E_0 C\gamma}{1 - LC\gamma^2}\sin\gamma t$$

$$= i_0\cos\frac{1}{\sqrt{LC}}t - \frac{1}{\sqrt{LC}}\left(q_0 - \frac{E_0 C}{1 - LC\gamma^2}\right)\sin\frac{1}{\sqrt{LC}}t - \frac{E_0 C\gamma}{1 - LC\gamma^2}\sin\gamma t.$$

5.2 | Linear Models: Boundary-Value Problems

The terminology and concepts listed below provide an outline of the main ideas encountered in this section. These can be useful when preparing for a quiz or test.

Terminology and Concepts

- axis of symmetry of a beam

- deflection curve or elastic curve of a beam

- cantilever beam

- eigenvalues and corresponding eigenfunctions of a boundary-value problem

The basic skills listed below summarize the more mechanical types of problems encountered in the exercise set for this section.

Basic Skills

- solve problems involving beams satisfying various end (boundary) conditions such as embedded, free, or simply supported

- find the eigenvalues and corresponding eigenfunctions for second-order boundary-value problems

3. (a) The general solution is

$$y(x) = c_1 + c_2 x + c_3 x^2 + c_4 x^3 + \frac{w_0}{24EI}x^4.$$

The boundary conditions are $y(0) = 0$, $y'(0) = 0$, $y(L) = 0$, $y''(L) = 0$. The first two conditions give $c_1 = 0$ and $c_2 = 0$. The conditions at $x = L$ give the system

$$c_3 L^2 + c_4 L^3 + \frac{w_0}{24EI}L^4 = 0$$

$$2c_3 + 6c_4 L + \frac{w_0}{2EI} L^2 = 0.$$

Solving, we obtain $c_3 = w_0 L^2 / 16 EI$ and $c_4 = -5w_0 L / 48 EI$. The deflection is

$$y(x) = \frac{w_0}{48EI}(3L^2 x^2 - 5Lx^3 + 2x^4).$$

6. (a) $y_{\text{max}} = y(L) = w_0 L^4 / 8EI$

   (b) Replacing both L and x by $L/2$ in $y(x)$ we obtain $w_0 L^4 / 128 EI$, which is $1/16$ of the maximum deflection when the length of the beam is L.

   (c) $y_{\text{max}} = y(L/2) = 5w_0 L^4 / 384 EI$

   (d) The maximum deflection in Example 1 is $y(L/2) = (w_0 / 24EI)L^4 / 16 = w_0 L^4 / 384 EI$, which is $1/5$ of the maximum displacement of the beam in part (c).

9. This is Example 2 in the text with $L = \pi$. The eigenvalues are $\lambda_n = n^2 \pi^2 / \pi^2 = n^2$, $n = 1, 2, 3, \ldots$ and the corresponding eigenfunctions are $y_n = \sin(n\pi x / \pi) = \sin nx$, $n = 1, 2, 3, \ldots$.

12. For $\lambda \leq 0$ the only solution of the boundary-value problem is $y = 0$. For $\lambda = \alpha^2 > 0$ we have

$$y = c_1 \cos \alpha x + c_2 \sin \alpha x.$$

Since $y(0) = 0$ implies $c_1 = 0$, $y = c_2 \sin x \, dx$. Now

$$y'\left(\frac{\pi}{2}\right) = c_2 \alpha \cos \alpha \frac{\pi}{2} = 0$$

gives

$$\alpha \frac{\pi}{2} = \frac{(2n-1)\pi}{2} \quad \text{or} \quad \lambda = \alpha^2 = (2n-1)^2, \ n = 1, 2, 3, \ldots.$$

The eigenvalues $\lambda_n = (2n-1)^2$ correspond to the eigenfunctions $y_n = \sin(2n-1)x$.

15. The auxiliary equation has solutions

$$m = \frac{1}{2}\left(-2 \pm \sqrt{4 - 4(\lambda + 1)}\right) = -1 \pm \alpha.$$

For $\lambda = -\alpha^2 < 0$ we have

$$y = e^{-x}(c_1 \cosh \alpha x + c_2 \sinh \alpha x).$$

The boundary conditions imply

$$y(0) = c_1 = 0$$

$$y(5) = c_2 e^{-5} \sinh 5\alpha = 0$$

so $c_1 = c_2 = 0$ and the only solution of the boundary-value problem is $y = 0$.

For $\lambda = 0$ we have

$$y = c_1 e^{-x} + c_2 x e^{-x}$$

and the only solution of the boundary-value problem is $y = 0$.

For $\lambda = \alpha^2 > 0$ we have

$$y = e^{-x} \left(c_1 \cos \alpha x + c_2 \sin \alpha x \right).$$

Now $y(0) = 0$ implies $c_1 = 0$, so

$$y(5) = c_2 e^{-5} \sin 5\alpha = 0$$

gives

$$5\alpha = n\pi \quad \text{or} \quad \lambda = \alpha^2 = \frac{n^2 \pi^2}{25}, \quad n = 1, 2, 3, \dots.$$

The eigenvalues $\lambda_n = \dfrac{n^2 \pi^2}{25}$ correspond to the eigenfunctions $y_n = e^{-x} \sin \dfrac{n\pi}{5} x$ for $n = 1,\ 2,\ 3,\ \dots$.

18. For $\lambda = 0$ the general solution is $y = c_1 + c_2 \ln x$. Now $y' = c_2/x$, so $y'(e^{-1}) = c_2 e = 0$ implies $c_2 = 0$. Then $y = c_1$ and $y(1) = 0$ gives $c_1 = 0$. Thus $y(x) = 0$.

For $\lambda = -\alpha^2 < 0$, $y = c_1 x^{-\alpha} + c_2 x^{\alpha}$. The boundary conditions give $c_2 = c_1 e^{2\alpha}$ and $c_1 = 0$, so that $c_2 = 0$ and $y(x) = 0$.

For $\lambda = \alpha^2 > 0$, $y = c_1 \cos(\alpha \ln x) + c_2 \sin(\alpha \ln x)$. From $y(1) = 0$ we obtain $c_1 = 0$ and $y = c_2 \sin(\alpha \ln x)$. Now $y' = c_2 (\alpha/x) \cos(\alpha \ln x)$, so $y'(e^{-1}) = c_2 e \alpha \cos \alpha = 0$ implies $\cos \alpha = 0$ or $\alpha = (2n-1)\pi/2$ and $\lambda = \alpha^2 = (2n-1)^2 \pi^2 / 4$ for $n = 1,\ 2,\ 3,\ \dots$. The corresponding eigenfunctions are

$$y_n = \sin \left(\frac{2n-1}{2} \pi \ln x \right).$$

21. If restraints are put on the column at $x = L/4$, $x = L/2$, and $x = 3L/4$, then the critical load will be P_4.

24. (a) The boundary-value problem is

$$\frac{d^4y}{dx^4} + \lambda\frac{d^2y}{dx^2} = 0, \quad y(0) = 0, y''(0) = 0, \ y(L) = 0, y'(L) = 0,$$

where $\lambda = \alpha^2 = P/EI$. The solution of the differential equation is $y = c_1\cos\alpha x + c_2\sin\alpha x + c_3 x + c_4$ and the conditions $y(0) = 0$, $y''(0) = 0$ yield $c_1 = 0$ and $c_4 = 0$. Next, by applying $y(L) = 0$, $y'(L) = 0$ to $y = c_2\sin\alpha x + c_3 x$ we get the system of equations

$$c_2\sin\alpha L + c_3 L = 0$$

$$\alpha c_2\cos\alpha L + c_3 = 0.$$

To obtain nontrivial solutions c_2, c_3, we must have the determinant of the coefficients equal to zero:

$$\begin{vmatrix} \sin\alpha L & L \\ \alpha\cos\alpha L & 1 \end{vmatrix} = 0 \quad \text{or} \quad \tan\beta = \beta,$$

where $\beta = \alpha L$. If β_n denotes the positive roots of the last equation, then the eigenvalues are found from $\beta_n = \alpha_n L = \sqrt{\lambda_n}\,L$ or $\lambda_n = (\beta_n/L)^2$. From $\lambda = P/EI$ we see that the critical loads are $P_n = \beta_n^2 EI/L^2$. With the aid of a CAS we find that the first positive root of $\tan\beta = \beta$ is (approximately) $\beta_1 = 4.4934$, and so the Euler load is (approximately) $P_1 = 20.1907EI/L^2$. Finally, if we use $c_3 = -c_2\alpha\cos\alpha L$, then the deflection curves are

$$y_n(x) = c_2\sin\alpha_n x + c_3 x = c_2\left[\sin\left(\frac{\beta_n}{L}x\right) - \left(\frac{\beta_n}{L}\cos\beta_n\right)x\right].$$

(b) With $L = 1$ and c_2 appropriately chosen, the general shape of the first buckling mode,

$$y_1(x) = c_2\left[\sin\left(\frac{4.4934}{L}x\right) - \left(\frac{4.4934}{L}\cos(4.4934)\right)x\right],$$

is shown below.

27. The auxiliary equation is $m^2 + m = m(m + 1) = 0$ so that $u(r) = c_1 r^{-1} + c_2$. The boundary conditions $u(a) = u_0$ and $u(b) = u_1$ yield the system $c_1 a^{-1} + c_2 = u_0$, $c_1 b^{-1} + c_2 = u_1$. Solving gives

$$c_1 = \left(\frac{u_0 - u_1}{b - a}\right)ab \quad \text{and} \quad c_2 = \frac{u_1 b - u_0 a}{b - a}.$$

Thus

$$u(r) = \left(\frac{u_0 - u_1}{b - a}\right)\frac{ab}{r} + \frac{u_1 b - u_0 a}{b - a}.$$

30. The initial-value problem is

$$x'' + \frac{2}{m}x' + \frac{k}{m}x = 0, \quad x(0) = 0, \ x'(0) = v_0.$$

With $k = 10$, the auxiliary equation has roots $\gamma = -1/m \pm \sqrt{1 - 10m}/m$. Consider the three cases:

(*i*) $m = \frac{1}{10}$. The roots are $\gamma_1 = \gamma_2 = 10$ and the solution of the differential equation is $x(t) = c_1 e^{-10t} + c_2 t e^{-10t}$. The initial conditions imply $c_1 = 0$ and $c_2 = v_0$ and so $x(t) = v_0 t e^{-10t}$. The condition $x(1) = 0$ implies $v_0 e^{-10} = 0$ which is impossible because $v_0 \neq 0$.

(*ii*) $1 - 10m > 0$ or $0 < m < \frac{1}{10}$. The roots are

$$\gamma_1 = -\frac{1}{m} - \frac{1}{m}\sqrt{1 - 10m} \quad \text{and} \quad \gamma_2 = -\frac{1}{m} + \frac{1}{m}\sqrt{1 - 10m}$$

and the solution of the differential equation is $x(t) = c_1 e^{\gamma_1 t} + c_2 e^{\gamma_2 t}$. The initial conditions imply

$$c_1 + c_2 = 0$$

$$\gamma_1 c_1 + \gamma_2 c_2 = v_0$$

so $c_1 = v_0/(\gamma_1 - \gamma_2)$, $c_2 = -v_0/(\gamma_1 - \gamma_2)$, and

$$x(t) = \frac{v_0}{\gamma_1 - \gamma_2}(e^{\gamma_1 t} - e^{\gamma_2 t}).$$

Again, $x(1) = 0$ is impossible because $v_0 \neq 0$.

(*iii*) $1 - 10m < 0$ or $m > \frac{1}{10}$. The roots of the auxiliary equation are

$$\gamma_1 = -\frac{1}{m} - \frac{1}{m}\sqrt{10m - 1}\,i \quad \text{and} \quad \gamma_2 = -\frac{1}{m} + \frac{1}{m}\sqrt{10m - 1}\,i$$

and the solution of the differential equation is

$$x(t) = c_1 e^{-t/m}\cos\frac{1}{m}\sqrt{10m - 1}\,t + c_2 e^{-t/m}\sin\frac{1}{m}\sqrt{10m - 1}\,t.$$

The initial conditions imply $c_1 = 0$ and $c_2 = mv_0/\sqrt{10m - 1}$, so that

$$x(t) = \frac{mv_0}{\sqrt{10m - 1}}e^{-t/m}\sin\left(\frac{1}{m}\sqrt{10m - 1}\,t\right),$$

The condition $x(1) = 0$ implies

$$\frac{mv_0}{\sqrt{10m - 1}}e^{-1/m}\sin\frac{1}{m}\sqrt{10m - 1} = 0$$

$$\sin\frac{1}{m}\sqrt{10m - 1} = 0$$

$$\frac{1}{m}\sqrt{10m-1} = n\pi$$

$$\frac{10m-1}{m^2} = n^2\pi^2, \ n = 1, 2, 3, \ldots$$

$$(n^2\pi^2)m^2 - 10m + 1 = 0$$

$$m = \frac{10\sqrt{100 - 4n^2\pi^2}}{2n^2\pi^2} = \frac{5 \pm \sqrt{25 - n^2\pi^2}}{n^2\pi^2}.$$

Since m is real, $25 - n^2\pi^2 \geq 0$. If $25 - n^2\pi^2 = 0$, then $n^2 = 25/\pi^2$, and n is not an integer. Thus, $25 - n^2\pi^2 = (5 - n\pi)(5 + n\pi) > 0$ and since $n > 0$, $5 + n\pi > 0$, so $5 - n\pi > 0$ also. Then $n < 5/\pi$, and so $n = 1$. Therefore, the mass m will pass through the equilibrium position when $t = 1$ for

$$m_1 = \frac{5 + \sqrt{25 - \pi^2}}{\pi^2} \quad \text{and} \quad m_2 = \frac{5 - \sqrt{25 - \pi^2}}{\pi^2}.$$

33. (a) A solution curve has the same y-coordinate at both ends of the interval $[-\pi, \pi]$ and the tangent lines at the endpoints of the interval are parallel.

(b) For $\lambda = 0$ the solution of $y'' = 0$ is $y = c_1 x + c_2$. From the first boundary condition we have

$$y(-\pi) = -c_1\pi + c_2 = y(\pi) = c_1\pi + c_2$$

or $2c_1\pi = 0$. Thus, $c_1 = 0$ and $y = c_2$. This constant solution is seen to satisfy the boundary-value problem.

For $\lambda = -\alpha^2 < 0$ we have $y = c_1 \cosh \alpha x + c_2 \sinh \alpha x$. In this case the first boundary condition gives

$$y(-\pi) = c_1 \cosh(-\alpha\pi) + c_2 \sinh(-\alpha\pi)$$

$$= c_1 \cosh \alpha\pi - c_2 \sinh \alpha\pi$$

$$= y(\pi) = c_1 \cosh \alpha\pi + c_2 \sinh \alpha\pi$$

or $2c_2 \sinh \alpha\pi = 0$. Thus $c_2 = 0$ and $y = c_1 \cosh \alpha x$. The second boundary condition implies in a similar fashion that $c_1 = 0$. Thus, for $\lambda < 0$, the only solution of the boundary-value problem is $y = 0$.

For $\lambda = \alpha^2 > 0$ we have $y = c_1 \cos \alpha x + c_2 \sin \alpha x$. The first boundary condition implies

$$y(-\pi) = c_1 \cos(-\alpha\pi) + c_2 \sin(-\alpha\pi)$$

$$= c_1 \cos \alpha\pi - c_2 \sin \alpha\pi$$

$$= y(\pi) = c_1 \cos \alpha\pi + c_2 \sin \alpha\pi$$

or $2c_2 \sin \alpha \pi = 0$. Similarly, the second boundary condition implies $2c_1 \alpha \sin \alpha \pi = 0$. If $c_1 = c_2 = 0$ the solution is $y = 0$. However, if $c_1 \neq 0$ or $c_2 \neq 0$, then $\sin \alpha \pi = 0$, which implies that α must be an integer, n. Therefore, for c_1 and c_2 not both 0, $y = c_1 \cos nx + c_2 \sin nx$ is a nontrivial solution of the boundary-value problem. Since $\cos(-nx) = \cos nx$ and $\sin(-nx) = -\sin nx$, we may assume without loss of generality that the eigenvalues are $\lambda_n = \alpha^2 = n^2$, for n a positive integer. The corresponding eigenfunctions are $y_n = \cos nx$ and $y_n = \sin nx$.

(c)

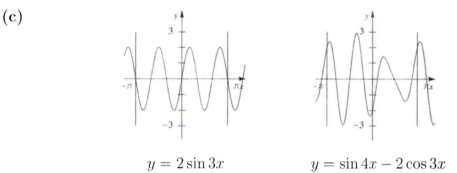

$$y = 2 \sin 3x \qquad\qquad y = \sin 4x - 2\cos 3x$$

36. Using a CAS we find that the first four nonnegative roots of $\tan x = -x$ are approximately $2.02876, 4.91318, 7.97867$, and 11.0855. The corresponding eigenvalues are $4.11586, 24.1393, 63.6591$, and 122.889, with eigenfunctions $\sin(2.02876x), \sin(4.91318x), \sin(7.97867x)$, and $\sin(11.0855x)$.

5.3 | Nonlinear Models

The terminology and concepts listed below provide an outline of the main ideas encountered in this section. These can be useful when preparing for a quiz or test.

Terminology and Concepts

- nonlinear spring models

- hard spring

- soft spring

- simple pendulum

- linearization of a DE

The basic skills listed below summarize the more mechanical types of problems encountered in the exercise set for this section.

Basic Skills

- use a numerical solver to plot solution curves of nonlinear DEs

Nonlinear Springs

3. The period corresponding to $x(0) = 1$, $x'(0) = 1$ is approximately 5.8. The second initial-value problem does not have a periodic solution.

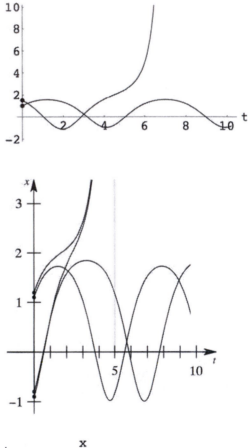

6. From the graphs we see that the interval is approximately $(-0.8, 1.1)$.

9. This is a damped hard spring, so x will approach 0 as t approaches ∞.

12. (a)

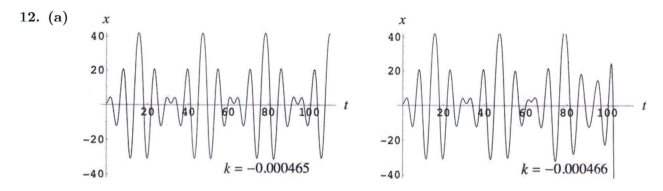

The system appears to be oscillatory for $-0.000465 \le k_1 < 0$ and nonoscillatory for $k_1 \le -0.000466$.

(b)

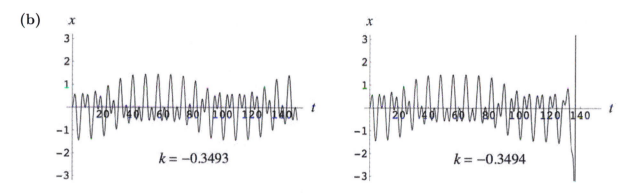

The system appears to be oscillatory for $-0.3493 \le k_1 < 0$ and nonoscillatory for $k_1 \le -0.3494$.

Variable Mass

15. (a) Intuitively, one might expect that only half of a 10-pound chain could be lifted by a 5-pound vertical force.

(b) Since $x = 0$ when $t = 0$, and $v = dx/dt = \sqrt{160 - 64x/3}$, we have $v(0) = \sqrt{160} \approx 12.65$ ft/s.

(c) Since x should always be positive, we solve $x(t) = 0$, getting $t = 0$ and $t = \frac{3}{2}\sqrt{5/2} \approx 2.3717$. Since the graph of $x(t)$ is a parabola, the maximum value occurs at $t_m = \frac{3}{4}\sqrt{5/2}$. This can also be obtained by solving $x'(t) = 0$. At this time the height of the chain is $x(t_m) \approx 7.5$ ft. This is higher than predicted because of the momentum generated by the force. When the chain is 5 feet high it still has a positive velocity of about 7.3 ft/s, which keeps it going higher for a while.

(d) As discussed in the solution to part **(c)** of this problem, the chain has momentum generated by the force applied to it that will cause it to go higher than expected. It will then fall back to below the expected maximum height, again due to momentum. This, in turn, will cause it to next go higher than expected, and so on.

Miscellaneous Mathematical Models

18. (a) Let (r, θ) denote the polar coordinates of the destroyer S_1. When S_1 travels the 6 miles from $(9, 0)$ to $(3, 0)$ it stands to reason, since S_2 travels half as fast as S_1, that the polar coordinates of S_2 are $(3, \theta_2)$, where θ_2 is unknown. In other words, the distances of the ships from $(0, 0)$ are the same and $r(t) = 15t$ then gives the radial distance of both ships. This is necessary if S_1 is to intercept S_2.

(b) The differential of arc length in polar coordinates is $(ds)^2 = (r\, d\theta)^2 + (dr)^2$, so that

$$\left(\frac{ds}{dt}\right)^2 = r^2 \left(\frac{d\theta}{dt}\right)^2 + \left(\frac{dr}{dt}\right)^2.$$

Using $ds/dt = 30$ and $dr/dt = 15$ then gives

$$900 = 225t^2 \left(\frac{d\theta}{dt}\right)^2 + 225$$

$$675 = 225t^2 \left(\frac{d\theta}{dt}\right)^2$$

$$\frac{d\theta}{dt} = \frac{\sqrt{3}}{t}$$

$$\theta(t) = \sqrt{3}\ln t + c = \sqrt{3}\ln \frac{r}{15} + c.$$

When $r = 3$, $\theta = 0$, so $c = -\sqrt{3}\ln\frac{1}{5}$ and

$$\theta(t) = \sqrt{3}\left(\ln \frac{r}{15} - \ln \frac{1}{5}\right) = \sqrt{3}\ln\frac{r}{3}.$$

Thus $r = 3e^{\theta/\sqrt{3}}$, whose graph is a logarithmic spiral.

(c) The time for S_1 to go from $(9, 0)$ to $(3, 0) = \frac{1}{5}$ hour. Now S_1 must intercept the path of S_2 for some angle β, where $0 < \beta < 2\pi$. At the time of interception t_2 we have $15t_2 = 3e^{\beta/\sqrt{3}}$ or $t = \frac{1}{5}e^{\beta/\sqrt{3}}$. The total time is then

$$t = \frac{1}{5} + \frac{1}{5}e^{\beta/\sqrt{3}} < \frac{1}{5}(1 + e^{2\pi/\sqrt{3}}).$$

5.R | Chapter 5 in Review

3. $5/4$ m, since $x = -\cos 4t + \frac{3}{4}\sin 4t$

6. False; since the equation of motion in this case is $x(t) = e^{-\lambda t}(c_1 + c_2 t)$ and $x(t) = 0$ can have at most one real solution

9. $y = 0$ because $\lambda = 8$ is not an eigenvalue

12. (a) Solving $\frac{3}{8}x'' + 6x = 0$ subject to $x(0) = 1$ and $x'(0) = -4$ we obtain

$$x = \cos 4t - \sin 4t = \sqrt{2}\sin(4t + 3\pi/4).$$

   (b) The amplitude is $\sqrt{2}$, period is $\pi/2$, and frequency is $2/\pi$.

   (c) If $x = 1$ then $t = n\pi/2$ and $t = -\pi/8 + n\pi/2$ for $n = 1, 2, 3, \ldots$.

   (d) If $x = 0$ then $t = \pi/16 + n\pi/4$ for $n = 0, 1, 2, \ldots$. The motion is upward for n even and downward for n odd.

   (e) $x'(3\pi/16) = 0$

   (f) If $x' = 0$ then $4t + 3\pi/4 = \pi/2 + n\pi$ or $t = 3\pi/16 + n\pi$.

15. From $mx'' + 4x' + 2x = 0$ we see that nonoscillatory motion results if $16 - 8m \geq 0$ or $0 < m \leq 2$.

18. Clearly $x_p = A/\omega^2$ suffices.

21. From $q'' + 10^4 q = 100\sin 50t$, $q(0) = 0$, and $q'(0) = 0$ we obtain $q_c = c_1\cos 100t + c_2\sin 100t$, $q_p = \frac{1}{75}\sin 50t$, and

   (a) $q = -\frac{1}{150}\sin 100t + \frac{1}{75}\sin 50t$,

   (b) $i = -\frac{2}{3}\cos 100t + \frac{2}{3}\cos 50t$, and

   (c) $q = 0$ when $\sin 50t(1 - \cos 50t) = 0$ or $t = n\pi/50$ for $n = 0, 1, 2, \ldots$.

24. (a) The differential equation is $d^2r/dt^2 - \omega^2 r = -g\sin\omega t$. The auxiliary equation is $m^2 - \omega^2 = 0$, so $r_c = c_1 e^{\omega t} + c_2 e^{-\omega t}$. A particular solution has the form $r_p = A\sin\omega t + B\cos\omega t$. Substituting into the differential equation we find $-2A\omega^2 \sin\omega t - 2B\omega^2 \cos\omega t = -g\sin\omega t$. Thus, $B = 0$, $A = g/2\omega^2$, and $r_p = (g/2\omega^2)\sin\omega t$. The general solution of the differential equation is $r(t) = c_1 e^{\omega t} + c_2 e^{-\omega t} + (g/2\omega^2)\sin\omega t$. The initial conditions imply $c_1 + c_2 = r_0$ and $g/2\omega - \omega c_1 + \omega c_2 = v_0$. Solving for c_1 and c_2 we get

$$c_1 = (2\omega^2 r_0 + 2\omega v_0 - g)/4\omega^2 \quad \text{and} \quad c_2 = (2\omega^2 r_0 - 2\omega v_0 + g)/4\omega^2,$$

so that

$$r(t) = \frac{2\omega^2 r_0 + 2\omega v_0 - g}{4\omega^2} e^{\omega t} + \frac{2\omega^2 r_0 - 2\omega v_0 + g}{4\omega^2} e^{-\omega t} + \frac{g}{2\omega^2} \sin\omega t.$$

(b) The bead will exhibit simple harmonic motion when the exponential terms are missing. Solving $c_1 = 0$, $c_2 = 0$ for r_0 and v_0 we find $r_0 = 0$ and $v_0 = g/2\omega$.

To find the minimum length of rod that will accommodate simple harmonic motion we determine the amplitude of $r(t)$ and double it. Thus $L = g/\omega^2$.

(c) As t increases, $e^{\omega t}$ approaches infinity and $e^{-\omega t}$ approaches 0. Since $\sin\omega t$ is bounded, the distance, $r(t)$, of the bead from the pivot point increases without bound and the distance of the bead from P will eventually exceed $L/2$.

(d)

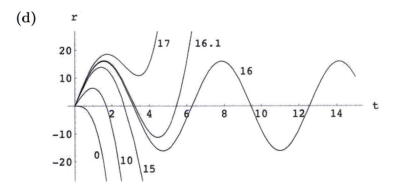

(e) For each v_0 we want to find the smallest value of t for which $r(t) = \pm 20$. Whether we look for $r(t) = -20$ or $r(t) = 20$ is determined by looking at the graphs in part (d). The total times that the bead stays on the rod is shown in the table below.

v_0	0	10	15	16.1	17
r	−20	−20	−20	20	20
t	1.55007	2.35494	3.43088	6.11627	4.22339

When $v_0 = 16$ the bead never leaves the rod.

27. The force of kinetic friction opposing the motion of the mass in μN, where μ is the coefficient of sliding friction and N is the normal component of the weight. Since friction is a force opposite to the direction of motion and since N is pointed directly downward (it is simply the weight of the mass), Newton's second law gives, for motion to the right $(x' > 0)$,

$$m\frac{d^2x}{dt^2} = -kx - \mu mg,$$

and for motion to the left $(x' < 0)$,

$$m\frac{d^2x}{dt^2} = -kx + \mu mg.$$

Traditionally, these two equations are written as one expression

$$m\frac{d^2x}{dt^2} + f_k\,\text{sgn}(x') + kx = 0,$$

where $f_k = \mu mg$ and

$$\text{sgn}(x') = \begin{cases} 1, & x' > 0 \\ -1, & x' < 0. \end{cases}$$

6 | Series Solutions of Linear Equations

6.1 | Review of Power Series

The terminology and concepts listed below provide an outline of the main ideas encountered in this section. These can be useful when preparing for a quiz or test.

Terminology and Concepts

- power series centered at a

- Maclaurin series

- interval and radius of convergence

- absolute convergence

- ratio test

- arithmetic of power series

- solution of differential equations using power series

Basic Skills

- find the interval and radius of convergence of a power series

- use known power series find new power series

- be able to shift the summation index of a power series

- be able to solve simple first-order, linear differential equations

3. $\lim\limits_{n\to\infty} \left| \dfrac{2^{n+1}x^{n+1}/(n+1)}{2^n x^n/n} \right| = \lim\limits_{n\to\infty} \dfrac{2n}{n+1}|x| = 2|x|$

The series is absolutely convergent for $2|x| < 1$ or $|x| < \frac{1}{2}$. The radius of convergence is $R = \frac{1}{2}$.

At $x = -\frac{1}{2}$, the series $\sum\limits_{n=1}^{\infty}(-1)^n/n$ converges by the alternating series test. At $x = \frac{1}{2}$, the series

$\sum\limits_{n=1}^{\infty} 1/n$ is the harmonic series which diverges. Thus, the given series converges on $[-\frac{1}{2}, \frac{1}{2})$, and

the radius of convergence is $\frac{1}{2}$.

6. $\lim\limits_{k\to\infty} \left| \dfrac{(k+1)!(x-1)^{k+1}}{k!(x-1)^k} \right| = \lim\limits_{k\to\infty} (k+1)|x-1| = \begin{cases} \infty, & x \neq 1 \\ 0, & x = 1 \end{cases}$

The radius of convergence is $R = 0$ and the series converges only for $x = 1$.

9. Write the series as $\sum\limits_{k=1}^{\infty} \left(\dfrac{32}{75}\right)^k x^k$. Then

$$\lim\limits_{n\to\infty} \left| \dfrac{a_{n+1}}{a_n} \right| = \lim\limits_{n\to\infty} \left| \dfrac{(32/75)^{n+1}x^{n+1}}{(32/75)^n x^n} \right| = \lim\limits_{n\to\infty} \left| \dfrac{32}{75}x \right| = \dfrac{32}{75}|x|.$$

The series is absolutely convergent for $\dfrac{32}{75}|x| < 1$, or on $(-75/32, 75/32)$. At $x = -75/32$ the

series $\sum\limits_{k=1}^{\infty}(-1)^k$ diverges by the n-th term test. At $x = 75/32$ the series $\sum\limits_{k=1}^{\infty} 1$ diverges by

the n-th term test. Thus, the given series converges on $(-75/32, 75/32)$, and the radius of

convergence is $75/32$.

12. We replace x by $3x$ in the Maclaurin series of e^x and multiply the result by x.

$$xe^{3x} = x \cdot \sum_{n=0}^{\infty} \frac{1}{n!}(3x)^n = \sum_{n=0}^{\infty} \frac{3^n}{n!}x^{n+1}$$

15. We replace x by $-x$ in the Maclaurin series of $\ln(1+x)$.

$$\ln(1-x) = -x - \frac{(-x)^2}{2} + \frac{(-x)^3}{3} + \cdots = -x - \frac{x^2}{2} - \frac{x^3}{3} - \cdots = \sum_{n=1}^{\infty} \frac{-1}{n}x^n$$

18. We first note that

$$\ln x = \ln\left[2\left(1 + \frac{x-2}{2}\right)\right] = \ln 2 + \ln\left(1 + \frac{x-2}{2}\right).$$

Next use the Maclaurin series of $\ln(1+x)$ with x replaced by $\dfrac{x-2}{2}$.

$$\ln x = \ln 2 + \frac{x-2}{2} - \frac{1}{2}\left(\frac{x-2}{2}\right)^2 + \frac{1}{3}\left(\frac{x-2}{2}\right)^3 - \frac{1}{4}\left(\frac{x-2}{2}\right)^4 + \cdots$$

$$= \ln 2 + \sum_{n=1}^{\infty} \frac{(-1)^{n+1}}{n2^n}(x-2)^n$$

21. $\sec x = \dfrac{1}{\cos x} = \dfrac{1}{1 - \dfrac{x^2}{2} + \dfrac{x^4}{4!} - \dfrac{x^6}{6!} + \cdots} = 1 + \dfrac{x^2}{2} + \dfrac{5x^4}{4!} + \dfrac{61x^6}{6!} + \cdots$

Since $\cos(\pi/2) = \cos(-\pi/2) = 0$, the series converges on $(-\pi/2, \pi/2)$.

24. Let $k = n - 3$ so that $n = k + 3$ and

$$\sum_{n=3}^{\infty}(2n-1)c_n x^{n-3} = \sum_{k=0}^{\infty}(2k+5)c_{k+3}x^k.$$

27. In the first summation let $k = n-1$ and in the second summation let $k = n+1$. Then $n = k+1$ in the first summation, $n = k - 1$ in the second summation, and

$$\sum_{n=1}^{\infty} 2nc_n x^{n-1} + \sum_{n=0}^{\infty} 6c_n x^{n+1} = 2 \cdot 1 \cdot c_1 x^0 + \sum_{n=2}^{\infty} 2nc_n x^{n-1} + \sum_{n=0}^{\infty} 6c_n x^{n+1}$$

$$= 2c_1 + \sum_{k=1}^{\infty} 2(k+1)c_{k+1}x^k + \sum_{k=1}^{\infty} 6c_{k-1}x^k$$

$$= 2c_1 + \sum_{k=1}^{\infty} \left[2(k+1)c_{k+1} + 6c_{k-1}\right]x^k.$$

30. In the first and third summations let $k = n$ and in the second summation let $k = n - 2$. Then $n = k$ in the first and third summations, $n = k + 2$ in the second summation, and

$$\sum_{n=2}^{\infty} n(n-1)c_n x^n + 2\sum_{n=2}^{\infty} n(n-1)c_n x^{n-2} + 3\sum_{n=1}^{\infty} nc_n x^n$$

$$= 2\cdot 2 \cdot 1 c_2 x^0 + 2\cdot 3 \cdot 2 c_3 x^1 + 3\cdot 1 \cdot c_1 x^1 + \sum_{n=2}^{\infty} n(n-1)c_n x^n + 2\sum_{n=4}^{\infty} n(n-1)c_n x^{n-2} + 3\sum_{n=2}^{\infty} nc_n x^n$$

$$= 4c_2 + (3c_1 + 12c_3)x + \sum_{k=2}^{\infty} k(k-1)c_k x^k + 2\sum_{k=2}^{\infty}(k+2)(k+1)c_{k+2}x^k + 3\sum_{k=2}^{\infty} kc_k x^k$$

$$= 4c_2 + (3c_1 + 12c_3)x + \sum_{k=2}^{\infty} \left[(k(k-1)+3k)c_k + 2(k+2)(k+1)c_{k+2}\right]x^k$$

$$= 4c_2 + (3c_1 + 12c_3)x + \sum_{k=2}^{\infty} \left[k(k+2)c_k + 2(k+1)(k+2)c_{k+2}\right]x^k$$

33. In this problem we must take special care with starting values for the indices of summation. Normally when a power series is given in summation notation, successive derivatives of the power series of the unknown function start with an index that is one higher than the preceding one. In this case, the power series starts with $n = 1$ to avoid division by zero. From the first derivative on this is no longer necessary and the index of summation starts again with $n = 1$ for y'. To justify this to yourself you could simply write out the first few term of the power series for y.

Since

$$y' = \sum_{n=1}^{\infty} (-1)^{n+1} x^{n-1} \quad \text{and} \quad y'' = \sum_{n=2}^{\infty} (-1)^{n+1}(n-1)x^{n-2},$$

we have

$$(x+1)y'' + y' = (x+1)\sum_{n=2}^{\infty}(-1)^{n+1}(n-1)x^{n-2} + \sum_{n=1}^{\infty}(-1)^{n+1}x^{n-1}$$

$$= \sum_{n=2}^{\infty}(-1)^{n+1}(n-1)x^{n-1} + \sum_{n=2}^{\infty}(-1)^{n+1}(n-1)x^{n-2} + \sum_{n=1}^{\infty}(-1)^{n+1}x^{n-1}$$

$$= \underbrace{\sum_{n=2}^{\infty}(-1)^{n+1}(n-1)x^{n-1}}_{k=n-1} + \underbrace{\sum_{n=2}^{\infty}(-1)^{n+1}(n-1)x^{n-2}}_{k=n-2} + \underbrace{\sum_{n=1}^{\infty}(-1)^{n+1}x^{n-1}}_{k=n-1}$$

$$= \sum_{k=1}^{\infty}(-1)^{k+2}kx^{k} + \sum_{k=0}^{\infty}(-1)^{k+3}(k+1)x^{k} + \sum_{k=0}^{\infty}(-1)^{k+2}x^{k}$$

$$= -x^0 + x^0 + \sum_{k=2}^{\infty}\left[(-1)^{k+2}k - (-1)^{k+2}k - (-1)^{k+2} + (-1)^{k+2}\right]x^{k} = 0.$$

36. Substituting into the differential equation we have

$$4y' + y = 4\sum_{n=1}^{\infty} nc_n x^{n-1} + \sum_{n=0}^{\infty} c_n x^n = 4\sum_{k=0}^{\infty}(k+1)c_{k+1}x^k + \sum_{k=0}^{\infty} c_k x^k$$

$$= \sum_{k=0}^{\infty}\left[4(k+1)c_{k+1} + c_k\right]x^k = 0.$$

Thus $c_{k+1} = -\dfrac{1}{4(k+1)} c_k,$ for $k = 0, 1, 2, \ldots,$ and

$$c_1 = -\frac{1}{4 \cdot 1} c_0$$

$$c_2 = -\frac{1}{4 \cdot 2} c_1 = \frac{1}{4^2 \cdot 1 \cdot 2} c_0$$

$$c_3 = -\frac{1}{4 \cdot 3} c_2 = -\frac{1}{4^3 \cdot 1 \cdot 2 \cdot 3} c_0$$

$$c_4 = -\frac{1}{4 \cdot 4} c_3 = \frac{1}{4^4 \cdot 1 \cdot 2 \cdot 3 \cdot 4} c_0$$

$$\vdots$$

Hence,

$$y = c_0 - \frac{1}{4 \cdot 1} c_0 x + \frac{1}{4^2 \cdot 1 \cdot 2} c_0 x^2 - \frac{1}{4^3 \cdot 1 \cdot 2 \cdot 3} c_0 x^3 + \frac{1}{4^4 \cdot 1 \cdot 2 \cdot 3 \cdot 4} c_0 x^4 - \cdots$$

and

$$y = c_0 \sum_{k=0}^{\infty} \frac{(-1)^k}{4^k k!} x^k = c_0 \sum_{k=0}^{\infty} \frac{(-1^k}{k!} \left(\frac{x}{4}\right)^k = c_0 e^{-x/4}.$$

6.2 | Solutions About Ordinary Points

The terminology and concepts listed below provide an outline of the main ideas encountered in this section. These can be useful when preparing for a quiz or test.

Terminology and Concepts

- power series centered at a; $\sum_{n=0}^{\infty} c_n (x-a)^n$

- convergence of a power series

- interval of convergence

- radius of convergence

- absolute convergence

- ratio test

- a power series defines a function

- a function analytic at a point

- ordinary point

- singular point

- power series solution of a DE

- recurrence relation

The basic skills listed below summarize the more mechanical types of problems encountered in the exercise set for this section.

Basic Skills

- find the radius of convergence of a power series

- convert an expression like $\sum_{n=3}^{\infty}(n+1)x^{n-2} + \sum_{n=2}^{\infty} nx^n$ to an expression involving a single infinite summation: $4x + \sum_{k=2}^{\infty}(2k+3)x^k$

- find power series solutions of linear second-order DEs

Summation Notation

Without a doubt, the key to successfully solving a linear DE with variable coefficients by the power series method depends on your ability to recast a combination of power series centered at the same point in terms of a *single* summation. In Example 1 of Section 6.1 we gave an example involving two series; here is another example of how this is done using three series. We wish to combine the three sums

$$\sum_{n=2}^{\infty} c_n n(n-1)x^{n-2} + 2\sum_{n=1}^{\infty} c_n n x^n + 5\sum_{n=0}^{\infty} c_n x^{n+1}$$

into one term. Notice that the first series starts (for $n=2$) with x^0, the second series starts (for $n=1$) with x^1, and the third series also starts (for $n=0$) with x^1. To bring the first series in line with the last two, we write out the term corresponding to $n=2$ and observe that now all three series

$$2c_2 + \sum_{n=3}^{\infty} c_n n(n-1)x^{n-2} + \sum_{n=1}^{\infty} 2c_n n x^n + \sum_{n=0}^{\infty} 5c_n x^{n+1}$$

begin with the same power x^1. The next thing that must be done is to get a common summation index for the series:

- In the first series, we let $k = n - 2$. As $n = 3, 4, 5, \ldots,$ k takes on the values $1, 2, 3, \ldots$.

- In the second series, we let $k = n$. Here $n = 1, 2, 3, \ldots$ corresponds to $k = 1, 2, 3, \ldots$.

- In the third series, we let $k = n + 1$. Then $n = 0, 1, 2, \ldots$ also yields $k = 1, 2, 3, \ldots$.

Next, we substitute $n = k + 2$ in the first series, $n = k$ in the second series, and then $n = k - 1$ in the third:

$$2c_2 + \sum_{k=1}^{\infty} c_{k+2}(k+2)(k+1)x^k + \sum_{k=1}^{\infty} 2c_k k x^k + \sum_{k=1}^{\infty} 5c_{k-1}x^k.$$

At this point, when all the series start with the same power of x and have the same summation index starting with the same value (in this case, $k = 1$), we can add the series term-by-term to get

$$2c_2 + \sum_{k=1}^{\infty} [c_{k+2}(k+2)(k+1) + 2c_k k + 5c_{k-1}]x^k.$$

Alternatively, we can make the change of summation variable first as illustrated in the solution of Problem 18 in this section of the manual.

Use of Computers   Both *Mathematica* and *Maple* can express the sum of several terms in a concise way without actually listing each term. The commands below can be used to express the first four terms of the infinite sum $\sum_{n=0}^{\infty} c_n (-1)^n/(n+1)^2$.

Clear[c]	(*Mathematica*)
Sum[c[n](-1)^n/(n+1)^2, {n, 0, 3}]	

`sum(c(n)*(-1)^n/(n+1)^2, n=0..3);`	(*Maple*)

To obtain the power series representation about $x = 0$ of a function like $f(x) = e^x$ use

Series[Exp[x], {x, 0, 5}]	(*Mathematica*)

or

`series(exp(x),x=0,5);`	(*Maple*)

The *Mathematica* output corresponding to the above command is

$$1 + x + \frac{1}{2}x^2 + \frac{1}{6}x^3 + \frac{1}{24}x^4 + \frac{1}{120}x^5 + O[x]^6,$$

whereas for *Maple* the output is

$$1 + x + \frac{1}{2}x^2 + \frac{1}{6}x^3 + \frac{1}{24}x^4 + O(x^5).$$

Note that in the *Mathematica* command the 5 tells the program to compute the series out to, and including, the term containing x^5. On the other hand, in *Maple*, the 5 tells the program to compute the series out to, but excluding, the term containing x^5.

Depending on how you want to use the series representation of a function found using a CAS, it may be inconvenient to have the order of the series included in the output. For example, neither *Mathematica* nor *Maple* will plot the graph of an expression containing $O[x]\hat{\ }6$, in the case of *Mathematica*, or $O(x^5)$ in the case of *Maple*. To get around this, both programs have commands that truncate the series, converting it to a polynomial.

$$\textbf{Normal[Series[Exp[x], \{x, 0, 5\}]]} \hspace{3cm} (\textit{Mathematica})$$

or

$$\textbf{Series[Exp[x], \{x, 0, 5\}]//Normal}$$

$$\texttt{convert(series(exp(x),x=0,5),polynom);} \hspace{1.5cm} (\textit{Maple})$$

3. We have

$$y'' + y = \underbrace{\sum_{n=2}^{\infty} n(n-1)c_n x^{n-2}}_{k=n-2} + \underbrace{\sum_{n=0}^{\infty} c_n x^n}_{k=n}$$

$$= \sum_{k=0}^{\infty} (k+2)(k+1)c_{k+2}x^k + \sum_{k=0}^{\infty} c_k x^k$$

$$= \sum_{k=0}^{\infty} \left[(k+2)(k+1)c_{k+2} + c_k \right] x^k = 0.$$

Thus $c_{k+2} = -\dfrac{c_k}{(k+2)(k+1)}$, for $k = 0, 1, 2, \ldots$, and for $k = 0, 2, 4, 6, \ldots$ we get

$$c_2 = -\frac{c_0}{2!}$$

$$c_4 = \frac{c_0}{4!}$$

$$c_6 = -\frac{c_0}{6!}$$

$$\vdots$$

For $k = 1, 3, 5, 7, \ldots$ we get

$$c_3 = -\frac{c_1}{3!}$$

$$c_5 = \frac{c_1}{5!}$$

$$c_7 = -\frac{c_1}{7!}$$

$$\vdots$$

Hence,

$$y_1(x) = c_0 \left[1 - \frac{1}{2!}x^2 + \frac{1}{4!}x^4 - \frac{1}{6!}x^6 + \cdots \right]$$

and

$$y_2(x) = c_1 \left[x - \frac{1}{3!}x^3 + \frac{1}{5!}x^5 - \frac{1}{7!}x^7 + \cdots \right].$$

The solution $y_1(x)$ is recognized as $y_1(x) = c_0 \cos x$, and the solution $y_2(x)$ is recognized as $y_2(x) = c_1 \sin x$

6. We have

$$y'' + 2y' = \underbrace{\sum_{n=2}^{\infty} n(n-1)c_n x^{n-2}}_{k=n-2} + 2 \underbrace{\sum_{n=1}^{\infty} nc_n x^{n-1}}_{k=n-1}$$

$$= \sum_{k=0}^{\infty} (k+2)(k+1)c_{k+2} x^k + \sum_{k=0}^{\infty} 2(k+1)c_{k+1} x^k$$

$$= \sum_{k=0}^{\infty} \left[(k+2)(k+1)c_{k+2} + 2(k+1)c_{k+1} \right] x^k = 0.$$

Thus $c_{k+2} = -\dfrac{2(k+1)c_{k+1}}{(k+2)(k+1)} = -\dfrac{2c_{k+1}}{k+2}$, for $k = 0, 1, 2, \ldots,$ so

$$c_2 = -\frac{2c_1}{2!}$$

$$c_3 = \frac{2^2 c_2}{3} = \frac{2^2 c_1}{3!}$$

$$c_4 = -\frac{2^3 c_3}{4} = -\frac{2^3 c_1}{4!}$$

$$\vdots$$

Hence, the solution of the differential equation is

$$y(x) = y_1(x) + y_2(x) = c_0 + c_1 \left[x - \frac{2}{2!}x^2 + \frac{2^2}{3!}x^3 - \frac{2^3}{4!}x^4 + \cdots \right]$$

$$= c_0 + \frac{1}{2}c_1 \left[2x - \frac{2^2}{2!}x^2 + \frac{2^3}{3!}x^3 - \frac{2^4}{4!}x^4 + \cdots \right]$$

$$= c_0 + \frac{1}{2}c_1 \left[1 - 12x - \frac{2^2}{2!}x^2 + \frac{2^3}{3!}x^3 - \frac{2^4}{4!}x^4 + \cdots \right]$$

$$= c_0 + \frac{1}{2}c_1 - \frac{1}{2}c_1 \sum_{k=0}^{\infty} \frac{1}{k!}(2x)^k.$$

The solutions $y_1(x)$ and $y_2(x)$ are recognized as

$$y_1(x) = c_0 \qquad \text{and} \qquad y_2(x) = \frac{1}{2}c_1 - \frac{1}{2}c_1 e^{-2x}.$$

9. Substituting $y = \sum_{n=0}^{\infty} c_n x^n$ into the differential equation we have

$$y'' - 2xy' + y = \underbrace{\sum_{n=2}^{\infty} n(n-1)c_n x^{n-2}}_{k=n-2} - 2\underbrace{\sum_{n=1}^{\infty} nc_n x^n}_{k=n} + \underbrace{\sum_{n=0}^{\infty} c_n x^n}_{k=n}$$

$$= \sum_{k=0}^{\infty} (k+2)(k+1)c_{k+2}x^k - 2\sum_{k=1}^{\infty} kc_k x^k + \sum_{k=0}^{\infty} c_k x^k$$

$$= 2c_2 + c_0 + \sum_{k=1}^{\infty} [(k+2)(k+1)c_{k+2} - (2k-1)c_k]x^k = 0.$$

Thus

$$2c_2 + c_0 = 0$$

$$(k+2)(k+1)c_{k+2} - (2k-1)c_k = 0$$

and

$$c_2 = -\frac{1}{2}c_0$$

$$c_{k+2} = \frac{2k-1}{(k+2)(k+1)}c_k, \quad k = 1, 2, 3, \ldots .$$

Choosing $c_0 = 1$ and $c_1 = 0$ we find

$$c_2 = -\frac{1}{2}$$

$$c_3 = c_5 = c_7 = \cdots = 0$$

$$c_4 = -\frac{1}{8}$$

$$c_6 = -\frac{7}{240}$$

and so on. For $c_0 = 0$ and $c_1 = 1$ we obtain

$$c_2 = c_4 = c_6 = \cdots = 0$$

$$c_3 = \frac{1}{6}$$

$$c_5 = \frac{1}{24}$$

$$c_7 = \frac{1}{112}$$

and so on. Thus, two solutions are

$$y_1 = 1 - \frac{1}{2}x^2 - \frac{1}{8}x^4 - \frac{7}{240}x^6 - \cdots \quad \text{and} \quad y_2 = x + \frac{1}{6}x^3 + \frac{1}{24}x^5 + \frac{1}{112}x^7 + \cdots .$$

12. Substituting $y = \sum_{n=0}^{\infty} c_n x^n$ into the differential equation we have

$$y'' + 2xy' + 2y = \underbrace{\sum_{n=2}^{\infty} n(n-1)c_n x^{n-2}}_{k=n-2} + 2\underbrace{\sum_{n=1}^{\infty} nc_n x^n}_{k=n} + 2\underbrace{\sum_{n=0}^{\infty} c_n x^n}_{k=n}$$

$$= \sum_{k=0}^{\infty} (k+2)(k+1)c_{k+2} x^k + 2\sum_{k=1}^{\infty} kc_k x^k + 2\sum_{k=0}^{\infty} c_k x^k$$

$$= 2c_2 + 2c_0 + \sum_{k=1}^{\infty} [(k+2)(k+1)c_{k+2} + 2(k+1)c_k] x^k = 0.$$

Thus

$$2c_2 + 2c_0 = 0$$

$$(k+2)(k+1)c_{k+2} + 2(k+1)c_k = 0$$

and

$$c_2 = -c_0$$

$$c_{k+2} = -\frac{2}{k+2} c_k, \quad k = 1, 2, 3, \dots .$$

Choosing $c_0 = 1$ and $c_1 = 0$ we find

$$c_2 = -1$$

$$c_3 = c_5 = c_7 = \cdots = 0$$

$$c_4 = \frac{1}{2}$$

$$c_6 = -\frac{1}{6}$$

and so on. For $c_0 = 0$ and $c_1 = 1$ we obtain

$$c_2 = c_4 = c_6 = \cdots = 0$$

$$c_3 = -\frac{2}{3}$$

$$c_5 = \frac{4}{15}$$

$$c_7 = -\frac{8}{105}$$

and so on. Thus, two solutions are

$$y_1 = 1 - x^2 + \frac{1}{2}x^4 - \frac{1}{6}x^6 + \cdots \quad \text{and} \quad y_2 = x - \frac{2}{3}x^3 + \frac{4}{15}x^5 - \frac{8}{105}x^7 + \cdots .$$

15. Substituting $y = \sum_{n=0}^{\infty} c_n x^n$ into the differential equation we have

$$y'' - (x+1)y' - y = \underbrace{\sum_{n=2}^{\infty} n(n-1)c_n x^{n-2}}_{k=n-2} - \underbrace{\sum_{n=1}^{\infty} nc_n x^n}_{k=n} - \underbrace{\sum_{n=1}^{\infty} nc_n x^{n-1}}_{k=n-1} - \underbrace{\sum_{n=0}^{\infty} c_n x^n}_{k=n}$$

$$= \sum_{k=0}^{\infty} (k+2)(k+1)c_{k+2}x^k - \sum_{k=1}^{\infty} kc_k x^k - \sum_{k=0}^{\infty} (k+1)c_{k+1}x^k - \sum_{k=0}^{\infty} c_k x^k$$

$$= 2c_2 - c_1 - c_0 + \sum_{k=1}^{\infty} [(k+2)(k+1)c_{k+2} - (k+1)c_{k+1} - (k+1)c_k]x^k = 0.$$

Thus

$$2c_2 - c_1 - c_0 = 0$$

$$(k+2)(k+1)c_{k+2} - (k+1)(c_{k+1} + c_k) = 0$$

and

$$c_2 = \frac{c_1 + c_0}{2}$$

$$c_{k+2} = \frac{c_{k+1} + c_k}{k+2}, \quad k = 1, 2, 3, \dots.$$

Choosing $c_0 = 1$ and $c_1 = 0$ we find

$$c_2 = \frac{1}{2}, \qquad c_3 = \frac{1}{6}, \qquad c_4 = \frac{1}{6},$$

and so on. For $c_0 = 0$ and $c_1 = 1$ we obtain

$$c_2 = \frac{1}{2}, \qquad c_3 = \frac{1}{2}, \qquad c_4 = \frac{1}{4},$$

and so on. Thus, two solutions are

$$y_1 = 1 + \frac{1}{2}x^2 + \frac{1}{6}x^3 + \frac{1}{6}x^4 + \cdots \qquad \text{and} \qquad y_2 = x + \frac{1}{2}x^2 + \frac{1}{2}x^3 + \frac{1}{4}x^4 + \cdots.$$

18. Substituting $y = \sum_{n=0}^{\infty} c_n x^n$ into the differential equation we have

$$(x^2-1)y'' + xy' - y = \underbrace{\sum_{n=2}^{\infty} n(n-1)c_n x^n}_{k=n} - \underbrace{\sum_{n=2}^{\infty} n(n-1)c_n x^{n-2}}_{k=n-2} + \underbrace{\sum_{n=1}^{\infty} nc_n x^n}_{k=n} - \underbrace{\sum_{n=0}^{\infty} c_n x^n}_{k=n}$$

$$= \sum_{k=2}^{\infty} k(k-1)c_k x^k - \sum_{k=0}^{\infty} (k+2)(k+1)c_{k+2}x^k + \sum_{k=1}^{\infty} kc_k x^k - \sum_{k=0}^{\infty} c_k x^k$$

$$= (-2c_2 - c_0) - 6c_3 x + \sum_{k=2}^{\infty} \left[-(k+2)(k+1)c_{k+2} + (k^2-1)c_k\right]x^k = 0.$$

Thus

$$-2c_2 - c_0 = 0$$

$$-6c_3 = 0$$

$$-(k+2)(k+1)c_{k+2} + (k-1)(k+1)c_k = 0$$

and

$$c_2 = -\frac{1}{2}c_0$$

$$c_3 = 0$$

$$c_{k+2} = \frac{k-1}{k+2}c_k, \quad k = 2, 3, 4, \ldots.$$

Choosing $c_0 = 1$ and $c_1 = 0$ we find

$$c_2 = -\frac{1}{2}$$

$$c_3 = c_5 = c_7 = \cdots = 0$$

$$c_4 = -\frac{1}{8}$$

and so on. For $c_0 = 0$ and $c_1 = 1$ we obtain

$$c_2 = c_4 = c_6 = \cdots = 0$$

$$c_3 = c_5 = c_7 = \cdots = 0.$$

Thus, two solutions are

$$y_1 = 1 - \frac{1}{2}x^2 - \frac{1}{8}x^4 - \cdots \qquad \text{and} \qquad y_2 = x.$$

21. Substituting $y = \sum_{n=0}^{\infty} c_n x^n$ into the differential equation we have

$$y'' - 2xy' + 8y = \underbrace{\sum_{n=2}^{\infty} n(n-1)c_n x^{n-2}}_{k=n-2} - 2\underbrace{\sum_{n=1}^{\infty} nc_n x^n}_{k=n} + 8\underbrace{\sum_{n=0}^{\infty} c_n x^n}_{k=n}$$

$$= \sum_{k=0}^{\infty}(k+2)(k+1)c_{k+2}x^k - 2\sum_{k=1}^{\infty} kc_k x^k + 8\sum_{k=0}^{\infty} c_k x^k$$

$$= 2c_2 + 8c_0 + \sum_{k=1}^{\infty}[(k+2)(k+1)c_{k+2} + (8-2k)c_k]x^k = 0.$$

Thus

$$2c_2 + 8c_0 = 0$$

$$(k+2)(k+1)c_{k+2} + (8 - 2k)c_k = 0$$

and

$$c_2 = -4c_0$$

$$c_{k+2} = \frac{2(k-4)}{(k+2)(k+1)} c_k, \quad k = 1, 2, 3, \ldots.$$

Choosing $c_0 = 1$ and $c_1 = 0$ we find

$$c_2 = -4$$

$$c_3 = c_5 = c_7 = \cdots = 0$$

$$c_4 = \frac{4}{3}$$

$$c_6 = c_8 = c_{10} = \cdots = 0.$$

For $c_0 = 0$ and $c_1 = 1$ we obtain

$$c_2 = c_4 = c_6 = \cdots = 0$$

$$c_3 = -1$$

$$c_5 = \frac{1}{10}$$

and so on. Thus,

$$y = C_1 \left(1 - 4x^2 + \frac{4}{3}x^4 \right) + C_2 \left(x - x^3 + \frac{1}{10}x^5 + \cdots \right)$$

and

$$y' = C_1 \left(-8x + \frac{16}{3}x^3 \right) + C_2 \left(1 - 3x^2 + \frac{1}{2}x^4 + \cdots \right).$$

The initial conditions imply $C_1 = 3$ and $C_2 = 0$, so

$$y = 3 \left(1 - 4x^2 + \frac{4}{3}x^4 \right) = 3 - 12x^2 + 4x^4.$$

24. Substituting $y = \sum_{n=0}^{\infty} c_n x^n$ into the differential equation we have

$$y'' + e^x y' - y = \sum_{n=2}^{\infty} n(n-1)c_n x^{n-2}$$

$$+ \left(1 + x + \frac{1}{2}x^2 + \frac{1}{6}x^3 + \cdots\right)\left(c_1 + 2c_2 x + 3c_3 x^2 + 4c_4 x^3 + \cdots\right) - \sum_{n=0}^{\infty} c_n x^n$$

$$= \left[2c_2 + 6c_3 x + 12c_4 x^2 + 20c_5 x^3 + \cdots\right]$$

$$+ \left[c_1 + (2c_2 + c_1)x + \left(3c_3 + 2c_2 + \frac{1}{2}c_1\right)x^2 + \cdots\right] - \left[c_0 + c_1 x + c_2 x^2 + \cdots\right]$$

$$= (2c_2 + c_1 - c_0) + (6c_3 + 2c_2)x + \left(12c_4 + 3c_3 + c_2 + \frac{1}{2}c_1\right)x^2 + \cdots = 0.$$

Thus

$$2c_2 + c_1 - c_0 = 0$$

$$6c_3 + 2c_2 = 0$$

$$12c_4 + 3c_3 + c_2 + \frac{1}{2}c_1 = 0$$

and

$$c_2 = \frac{1}{2}c_0 - \frac{1}{2}c_1$$

$$c_3 = -\frac{1}{3}c_2$$

$$c_4 = -\frac{1}{4}c_3 + \frac{1}{12}c_2 - \frac{1}{24}c_1.$$

Choosing $c_0 = 1$ and $c_1 = 0$ we find

$$c_2 = \frac{1}{2}, \qquad c_3 = -\frac{1}{6}, \qquad c_4 = 0$$

and so on. For $c_0 = 0$ and $c_1 = 1$ we obtain

$$c_2 = -\frac{1}{2}, \qquad c_3 = \frac{1}{6}, \qquad c_4 = -\frac{1}{24}$$

and so on. Thus, two solutions are

$$y_1 = 1 + \frac{1}{2}x^2 - \frac{1}{6}x^3 + \cdots \qquad \text{and} \qquad y_2 = x - \frac{1}{2}x^2 + \frac{1}{6}x^3 - \frac{1}{24}x^4 + \cdots.$$

6.3 | Solutions About Singular Points

The terminology and concepts listed below provide an outline of the main ideas encountered in this section. These can be useful when preparing for a quiz or test.

Terminology and Concepts

- regular singular point of a DE
- irregular singular point of a DE
- method of Frobenius for finding a series solution about a regular singular point of a DE
- indicial equation

The basic skills listed below summarize the more mechanical types of problems encountered in the exercise set for this section.

Basic Skills

- determine the singular points of a linear DE and classify them as regular or irregular
- use the method of Frobenius to obtain two linearly independent solutions about $x = 0$ of a homogeneous linear DE with regular singular point $x = 0$ when the indicial roots
    1. do not differ by an integer, or
    2. differ by an integer.

3. Irregular singular point: $x = 3$; regular singular point: $x = -3$

6. Irregular singular point: $x = 5$; regular singular point: $x = 0$

9. Irregular singular point: $x = 0$; regular singular points: $x = 2$, ± 5

12. Writing the differential equation in the form

$$y'' + \frac{x+3}{x} y' + 7xy = 0$$

we see that $x_0 = 0$ is a regular singular point. Multiplying by x^2, the differential equation can be put in the form

$$x^2 y'' + x(x+3)y' + 7x^3 y = 0.$$

We identify $p(x) = x + 3$ and $q(x) = 7x^3$.

15. Substituting $y = \sum_{n=0}^{\infty} c_n x^{n+r}$ into the differential equation and collecting terms, we obtain

$$2xy'' - y' + 2y = \left(2r^2 - 3r\right)c_0 x^{r-1} + \sum_{k=1}^{\infty} [2(k+r-1)(k+r)c_k - (k+r)c_k + 2c_{k-1}]x^{k+r-1} = 0,$$

which implies

$$2r^2 - 3r = r(2r - 3) = 0$$

and

$$(k+r)(2k+2r-3)c_k + 2c_{k-1} = 0.$$

The indicial roots are $r = 0$ and $r = 3/2$. For $r = 0$ the recurrence relation is

$$c_k = -\frac{2c_{k-1}}{k(2k-3)}, \quad k = 1, 2, 3, \ldots,$$

and

$$c_1 = 2c_0, \qquad c_2 = -2c_0, \qquad c_3 = \frac{4}{9}c_0,$$

and so on. For $r = 3/2$ the recurrence relation is

$$c_k = -\frac{2c_{k-1}}{(2k+3)k}, \quad k = 1, 2, 3, \ldots,$$

and

$$c_1 = -\frac{2}{5}c_0, \qquad c_2 = \frac{2}{35}c_0, \qquad c_3 = -\frac{4}{945}c_0,$$

and so on. The general solution on $(0, \infty)$ is

$$y = C_1\left(1 + 2x - 2x^2 + \frac{4}{9}x^3 + \cdots\right) + C_2 x^{3/2}\left(1 - \frac{2}{5}x + \frac{2}{35}x^2 - \frac{4}{945}x^3 + \cdots\right).$$

18. Substituting $y = \sum_{n=0}^{\infty} c_n x^{n+r}$ into the differential equation and collecting terms, we obtain

$$2x^2 y'' - xy' + (x^2 + 1) y = (2r^2 - 3r + 1) c_0 x^r + (2r^2 + r) c_1 x^{r+1}$$

$$+ \sum_{k=2}^{\infty} [2(k+r)(k+r-1)c_k - (k+r)c_k + c_k + c_{k-2}]x^{k+r}$$

$$= 0,$$

which implies

$$2r^2 - 3r + 1 = (2r - 1)(r - 1) = 0,$$
$$(2r^2 + r) c_1 = 0,$$

and

$$[(k + r)(2k + 2r - 3) + 1]c_k + c_{k-2} = 0.$$

The indicial roots are $r = 1/2$ and $r = 1$, so $c_1 = 0$. For $r = 1/2$ the recurrence relation is

$$c_k = -\frac{c_{k-2}}{k(2k - 1)}, \quad k = 2, 3, 4, \ldots,$$

and

$$c_2 = -\frac{1}{6}c_0, \qquad c_3 = 0, \qquad c_4 = \frac{1}{168}c_0,$$

and so on. For $r = 1$ the recurrence relation is

$$c_k = -\frac{c_{k-2}}{k(2k + 1)}, \quad k = 2, 3, 4, \ldots,$$

and

$$c_2 = -\frac{1}{10}c_0, \qquad c_3 = 0, \qquad c_4 = \frac{1}{360}c_0,$$

and so on. The general solution on $(0, \infty)$ is

$$y = C_1 x^{1/2} \left(1 - \frac{1}{6}x^2 + \frac{1}{168}x^4 + \cdots\right) + C_2 x \left(1 - \frac{1}{10}x^2 + \frac{1}{360}x^4 + \cdots\right).$$

21. Substituting $y = \sum_{n=0}^{\infty} c_n x^{n+r}$ into the differential equation and collecting terms, we obtain

$$2xy'' - (3 + 2x)y' + y = (2r^2 - 5r) c_0 x^{r-1} + \sum_{k=1}^{\infty} [2(k+r)(k+r-1)c_k$$

$$- 3(k+r)c_k - 2(k + r - 1)c_{k-1} + c_{k-1}]x^{k+r-1}$$

$$= 0,$$

which implies

$$2r^2 - 5r = r(2r - 5) = 0$$

and

$$(k + r)(2k + 2r - 5)c_k - (2k + 2r - 3)c_{k-1} = 0.$$

The indicial roots are $r = 0$ and $r = 5/2$. For $r = 0$ the recurrence relation is

$$c_k = \frac{(2k - 3)c_{k-1}}{k(2k - 5)}, \quad k = 1, 2, 3, \ldots,$$

and

$$c_1 = \frac{1}{3}c_0, \qquad c_2 = -\frac{1}{6}c_0, \qquad c_3 = -\frac{1}{6}c_0,$$

and so on. For $r = 5/2$ the recurrence relation is

$$c_k = \frac{2(k + 1)c_{k-1}}{k(2k + 5)}, \quad k = 1, 2, 3, \ldots,$$

and

$$c_1 = \frac{4}{7}c_0, \qquad c_2 = \frac{4}{21}c_0, \qquad c_3 = \frac{32}{693}c_0,$$

and so on. The general solution on $(0, \infty)$ is

$$y = C_1 \left(1 + \frac{1}{3}x - \frac{1}{6}x^2 - \frac{1}{6}x^3 + \cdots \right) + C_2 x^{5/2} \left(1 + \frac{4}{7}x + \frac{4}{21}x^2 + \frac{32}{693}x^3 + \cdots \right).$$

24. Substituting $y = \sum_{n=0}^{\infty} c_n x^{n+r}$ into the differential equation and collecting terms, we obtain

$$2x^2 y'' + 3xy' + (2x - 1)y = \left(2r^2 + r - 1\right) c_0 x^r$$

$$+ \sum_{k=1}^{\infty} [2(k + r)(k + r - 1)c_k + 3(k + r)c_k - c_k + 2c_{k-1}]x^{k+r}$$

$$= 0,$$

which implies

$$2r^2 + r - 1 = (2r - 1)(r + 1) = 0$$

and

$$[(k + r)(2k + 2r + 1) - 1]c_k + 2c_{k-1} = 0.$$

The indicial roots are $r = -1$ and $r = 1/2$. For $r = -1$ the recurrence relation is

$$c_k = -\frac{2c_{k-1}}{k(2k - 3)}, \quad k = 1, 2, 3, \ldots,$$

and

$$c_1 = 2c_0, \qquad c_2 = -2c_0, \qquad c_3 = \frac{4}{9}c_0,$$

and so on. For $r = 1/2$ the recurrence relation is

$$c_k = -\frac{2c_{k-1}}{k(2k+3)}, \quad k = 1, 2, 3, \ldots,$$

and

$$c_1 = -\frac{2}{5}c_0, \qquad c_2 = \frac{2}{35}c_0, \qquad c_3 = -\frac{4}{945}c_0,$$

and so on. The general solution on $(0, \infty)$ is

$$y = C_1 x^{-1}\left(1 + 2x - 2x^2 + \frac{4}{9}x^3 + \cdots\right) + C_2 x^{1/2}\left(1 - \frac{2}{5}x + \frac{2}{35}x^2 - \frac{4}{945}x^3 + \cdots\right).$$

27. Substituting $y = \sum_{n=0}^{\infty} c_n x^{n+r}$ into the differential equation and collecting terms, we obtain

$$xy'' - xy' + y = \left(r^2 - r\right)c_0 x^{r-1} + \sum_{k=0}^{\infty}[(k+r+1)(k+r)c_{k+1} - (k+r)c_k + c_k]x^{k+r} = 0$$

which implies

$$r^2 - r = r(r - 1) = 0$$

and

$$(k+r+1)(k+r)c_{k+1} - (k+r-1)c_k = 0.$$

The indicial roots are $r_1 = 1$ and $r_2 = 0$. For $r_1 = 1$ the recurrence relation is

$$c_{k+1} = \frac{kc_k}{(k+2)(k+1)}, \quad k = 0, 1, 2, \ldots,$$

so $c_1 = c_2 = c_3 = \cdots = 0$ and one solution is $y_1 = c_0 x$. A second solution is

$$y_2 = x\int \frac{e^{-\int(-1)\,dx}}{x^2}\,dx = x\int \frac{e^x}{x^2}\,dx = x\int \frac{1}{x^2}\left(1 + x + \frac{1}{2}x^2 + \frac{1}{3!}x^3 + \cdots\right)dx$$

$$= x\int\left(\frac{1}{x^2} + \frac{1}{x} + \frac{1}{2} + \frac{1}{3!}x + \frac{1}{4!}x^2 + \cdots\right)dx = x\left[-\frac{1}{x} + \ln x + \frac{1}{2}x + \frac{1}{12}x^2 + \frac{1}{72}x^3 + \cdots\right]$$

$$= x\ln x - 1 + \frac{1}{2}x^2 + \frac{1}{12}x^3 + \frac{1}{72}x^4 + \cdots.$$

The general solution on $(0, \infty)$ is

$$y = C_1 x + C_2 y_2(x).$$

30. Substituting $y = \sum_{n=0}^{\infty} c_n x^{n+r}$ into the differential equation and collecting terms, we obtain

$$xy'' + y' + y = r^2 c_0 x^{r-1} + \sum_{k=1}^{\infty} [(k+r)(k+r-1)c_k + (k+r)c_k + c_{k-1}]x^{k+r-1} = 0$$

which implies $r^2 = 0$ and

$$(k+r)^2 c_k + c_{k-1} = 0.$$

The indicial roots are $r_1 = r_2 = 0$ and the recurrence relation is

$$c_k = -\frac{c_{k-1}}{k^2}, \quad k = 1, 2, 3, \ldots.$$

One solution is

$$y_1 = c_0 \left(1 - x + \frac{1}{2^2} x^2 - \frac{1}{(3!)^2} x^3 + \frac{1}{(4!)^2} x^4 - \cdots \right) = c_0 \sum_{n=0}^{\infty} \frac{(-1)^n}{(n!)^2} x^n.$$

A second solution is

$$y_2 = y_1 \int \frac{e^{-\int (1/x)dx}}{y_1^2}\, dx = y_1 \int \frac{dx}{x \left(1 - x + \frac{1}{4}x^2 - \frac{1}{36}x^3 + \cdots \right)^2}$$

$$= y_1 \int \frac{dx}{x \left(1 - 2x + \frac{3}{2}x^2 - \frac{5}{9}x^3 + \frac{35}{288}x^4 - \cdots \right)}$$

$$= y_1 \int \frac{1}{x} \left(1 + 2x + \frac{5}{2}x^2 + \frac{23}{9}x^3 + \frac{677}{288}x^4 + \cdots \right) dx$$

$$= y_1 \int \left(\frac{1}{x} + 2 + \frac{5}{2}x + \frac{23}{9}x^2 + \frac{677}{288}x^3 + \cdots \right) dx$$

$$= y_1 \left[\ln x + 2x + \frac{5}{4}x^2 + \frac{23}{27}x^3 + \frac{677}{1{,}152}x^4 + \cdots \right]$$

$$= y_1 \ln x + y_1 \left(2x + \frac{5}{4}x^2 + \frac{23}{27}x^3 + \frac{677}{1{,}152}x^4 + \cdots \right).$$

The general solution on $(0, \infty)$ is

$$y = C_1 y_1(x) + C_2 y_2(x).$$

33. (a) From $t = 1/x$ we have $dt/dx = -1/x^2 = -t^2$. Then

$$\frac{dy}{dx} = \frac{dy}{dt}\frac{dt}{dx} = -t^2\frac{dy}{dt}$$

and

$$\frac{d^2y}{dx^2} = \frac{d}{dx}\left(\frac{dy}{dx}\right) = \frac{d}{dx}\left(-t^2\frac{dy}{dt}\right) = -t^2\frac{d^2y}{dt^2}\frac{dt}{dx} - \frac{dy}{dt}\left(2t\frac{dt}{dx}\right) = t^4\frac{d^2y}{dt^2} + 2t^3\frac{dy}{dt}\ .$$

Now

$$x^4\frac{d^2y}{dx^2} + \lambda y = \frac{1}{t^4}\left(t^4\frac{d^2y}{dt^2} + 2t^3\frac{dy}{dt}\right) + \lambda y = \frac{d^2y}{dt^2} + \frac{2}{t}\frac{dy}{dt} + \lambda y = 0$$

becomes

$$t\frac{d^2y}{dt^2} + 2\frac{dy}{dt} + \lambda ty = 0.$$

(b) Substituting $y = \sum_{n=0}^{\infty} c_n t^{n+r}$ into the differential equation and collecting terms, we obtain

$$t\frac{d^2y}{dt^2} + 2\frac{dy}{dt} + \lambda ty = (r^2 + r)c_0 t^{r-1} + (r^2 + 3r + 2)c_1 t^r$$

$$+ \sum_{k=2}^{\infty}[(k+r)(k+r-1)c_k + 2(k+r)c_k + \lambda c_{k-2}]t^{k+r-1}$$

$$= 0,$$

which implies

$$r^2 + r = r(r+1) = 0,$$
$$(r^2 + 3r + 2)\, c_1 = 0,$$

and

$$(k+r)(k+r+1)c_k + \lambda c_{k-2} = 0.$$

The indicial roots are $r_1 = 0$ and $r_2 = -1$, so $c_1 = 0$. For $r_1 = 0$ the recurrence relation is

$$c_k = -\frac{\lambda c_{k-2}}{k(k+1)}\,, \quad k = 2, 3, 4, \ldots,$$

and

$$c_2 = -\frac{\lambda}{3!}c_0$$
$$c_3 = c_5 = c_7 = \cdots = 0$$
$$c_4 = \frac{\lambda^2}{5!}c_0$$
$$\vdots$$
$$c_{2n} = (-1)^n\frac{\lambda^n}{(2n+1)!}c_0.$$

For $r_2 = -1$ the recurrence relation is

$$c_k = -\frac{\lambda c_{k-2}}{k(k-1)}, \quad k = 2, 3, 4, \ldots,$$

and

$$c_2 = -\frac{\lambda}{2!} c_0$$

$$c_3 = c_5 = c_7 = \cdots = 0$$

$$c_4 = \frac{\lambda^2}{4!} c_0$$

$$\vdots$$

$$c_{2n} = (-1)^n \frac{\lambda^n}{(2n)!} c_0.$$

The general solution on $(0, \infty)$ is

$$y(t) = c_1 \sum_{n=0}^{\infty} \frac{(-1)^n}{(2n+1)!} (\sqrt{\lambda}\, t)^{2n} + c_2 t^{-1} \sum_{n=0}^{\infty} \frac{(-1)^n}{(2n)!} (\sqrt{\lambda}\, t)^{2n}$$

$$= \frac{1}{t} \left[C_1 \sum_{n=0}^{\infty} \frac{(-1)^n}{(2n+1)!} (\sqrt{\lambda}\, t)^{2n+1} + C_2 \sum_{n=0}^{\infty} \frac{(-1)^n}{(2n)!} (\sqrt{\lambda}\, t)^{2n} \right]$$

$$= \frac{1}{t} [C_1 \sin \sqrt{\lambda}\, t + C_2 \cos \sqrt{\lambda}\, t].$$

(c) Using $t = 1/x$, the solution of the original equation is

$$y(x) = C_1 x \sin \frac{\sqrt{\lambda}}{x} + C_2 x \cos \frac{\sqrt{\lambda}}{x}.$$

6.4 | Special Functions

The terminology and concepts listed below provide an outline of the main ideas encountered in this section. These can be useful when preparing for a quiz or test.

Terminology and Concepts

- Bessel's DE of order ν
- Bessel functions of the first kind: $J_\nu(x)$ and $J_{-\nu}(x)$
- Bessel functions of the second kind: $Y_\nu(x)$
- parametric Bessel equation of order ν
- modified Bessel equation of order ν

- modified Bessel function of the first kind: $I_\nu(x)$

- modified Bessel function of the second kind: $K_\nu(x)$

- aging spring

- Euler's constant: γ

- differential recurrence relation

- spherical Bessel functions

- Legendre's DE of order n

- Legendre polynomials

- Rodrigues' formula

The basic skills listed below summarize the more mechanical types of problems encountered in the exercise set for this section.

Basic Skills

- solve Bessel's DE of order ν

- solve DEs of the form $x^2 y'' + xy' + (\alpha^2 x^2 - \nu^2)y = 0$

- solve DEs of the form $y'' + [(1-2a)/x]y' + [b^2 c^2 x^{2c-2} + (a^2 - p^2 c^2)/x^2]y = 0, \ p \geq 0$

- verify various recurrence relations involving Bessel functions

- use a recurrence relation to generate Legendre polynomials

- use Rodrigues' formula to generate Legendre polynomials

Use of Computers   See Section 2.3 in this manual for the syntax used to obtain Bessel functions and Legendre polynomials in *Mathematica* and *Maple*.

Bessel's Equation

3. Since $\nu^2 = 25/4$ the general solution is $y = c_1 J_{5/2}(x) + c_2 J_{-5/2}(x)$.

6. Since $\nu^2 = 4$ the general solution is $y = c_1 J_2(x) + c_2 Y_2(x)$.

9. We identify $\alpha = 5$ and $\nu = \frac{2}{3}$. Then the general solution is $y = c_1 J_{2/3}(5x) + c_2 J_{-2/3}(5x)$.

12. If $y = \sqrt{x}\, v(x)$ then

$$y' = x^{1/2} v'(x) + \frac{1}{2} x^{-1/2} v(x)$$

$$y'' = x^{1/2} v''(x) + x^{-1/2} v'(x) - \frac{1}{4} x^{-3/2} v(x)$$

and

$$x^2 y'' + \left(\alpha^2 x^2 - \nu^2 + \frac{1}{4} \right) y = x^{5/2} v''(x) + x^{3/2} v'(x) - \frac{1}{4} x^{1/2} v(x) + \left(\alpha^2 x^2 - \nu^2 + \frac{1}{4} \right) x^{1/2} v(x)$$

$$= x^{5/2} v''(x) + x^{3/2} v'(x) + (\alpha^2 x^{5/2} - \nu^2 x^{1/2}) v(x) = 0.$$

Multiplying by $x^{-1/2}$ we obtain

$$x^2 v''(x) + x v'(x) + (\alpha^2 x^2 - \nu^2) v(x) = 0,$$

whose solution is $v(x) = c_1 J_\nu(\alpha x) + c_2 Y_\nu(\alpha x)$. Then $y = c_1 \sqrt{x} \, J_\nu(\alpha x) + c_2 \sqrt{x} \, Y_\nu(\alpha x)$.

15. Write the differential equation in the form $y'' - (1/x)y' + y = 0$. This is the form of (18) in the text with $a = 1$, $c = 1$, $b = 1$, and $p = 1$, so, by (19) in the text, the general solution is

$$y = x[c_1 J_1(x) + c_2 Y_1(x)].$$

18. Write the differential equation in the form $y'' + (4 + 1/4x^2)y = 0$. This is the form of (18) in the text with $a = \frac{1}{2}$, $c = 1$, $b = 2$, and $p = 0$, so, by (19) in the text, the general solution is

$$y = x^{1/2}[c_1 J_0(2x) + c_2 Y_0(2x)].$$

21. Using the fact that $i^2 = -1$, along with the definition of $J_\nu(x)$ in (7) in the text, we have

$$I_\nu(x) = i^{-\nu} J_\nu(ix) = i^{-\nu} \sum_{n=0}^{\infty} \frac{(-1)^n}{n! \Gamma(1 + \nu + n)} \left(\frac{ix}{2} \right)^{2n+\nu}$$

$$= \sum_{n=0}^{\infty} \frac{(-1)^n}{n! \Gamma(1 + \nu + n)} i^{2n+\nu-\nu} \left(\frac{x}{2} \right)^{2n+\nu}$$

$$= \sum_{n=0}^{\infty} \frac{(-1)^n}{n! \Gamma(1 + \nu + n)} (i^2)^n \left(\frac{x}{2} \right)^{2n+\nu}$$

$$= \sum_{n=0}^{\infty} \frac{(-1)^{2n}}{n! \Gamma(1 + \nu + n)} \left(\frac{x}{2} \right)^{2n+\nu}$$

$$= \sum_{n=0}^{\infty} \frac{1}{n! \Gamma(1 + \nu + n)} \left(\frac{x}{2} \right)^{2n+\nu},$$

which is a real function.

24. Write the differential equation in the form $y'' + (4/x)y' + (1 + 2/x^2)y = 0$. This is the form of (18) in the text with

$$1 - 2a = 4 \quad \Longrightarrow \quad a = -\frac{3}{2}$$

$$2c - 2 = 0 \quad \Longrightarrow \quad c = 1$$

$$b^2 c^2 = 1 \quad \Longrightarrow \quad b = 1$$

$$a^2 - p^2 c^2 = 2 \quad \Longrightarrow \quad p = \frac{1}{2}.$$

Then, by (19), (23), and (24) in the text,

$$y = x^{-3/2}[c_1 J_{1/2}(x) + c_2 J_{-1/2}(x)] = x^{-3/2}\left[c_1\sqrt{\frac{2}{\pi x}}\sin x + c_2\sqrt{\frac{2}{\pi x}}\cos x\right]$$

$$= C_1\frac{1}{x^2}\sin x + C_2\frac{1}{x^2}\cos x.$$

27. (a) The recurrence relation follows from

$$-\nu J_\nu(x) + x J_{\nu-1}(x) = -\sum_{n=0}^{\infty}\frac{(-1)^n\nu}{n!\Gamma(1+\nu+n)}\left(\frac{x}{2}\right)^{2n+\nu} + x\sum_{n=0}^{\infty}\frac{(-1)^n}{n!\Gamma(\nu+n)}\left(\frac{x}{2}\right)^{2n+\nu-1}$$

$$= -\sum_{n=0}^{\infty}\frac{(-1)^n\nu}{n!\Gamma(1+\nu+n)}\left(\frac{x}{2}\right)^{2n+\nu} + \sum_{n=0}^{\infty}\frac{(-1)^n(\nu+n)}{n!\Gamma(1+\nu+n)}\cdot 2\left(\frac{x}{2}\right)\left(\frac{x}{2}\right)^{2n+\nu-1}$$

$$= \sum_{n=0}^{\infty}\frac{(-1)^n(2n+\nu)}{n!\Gamma(1+\nu+n)}\left(\frac{x}{2}\right)^{2n+\nu} = x J_\nu'(x).$$

(b) The formula in part **(a)** is a linear first-order differential equation in $J_\nu(x)$. An integrating factor for this equation is x^ν, so

$$\frac{d}{dx}[x^\nu J_\nu(x)] = x^\nu J_{\nu-1}(x).$$

30. From (20) we obtain $J_0'(x) = -J_1(x)$, and from (21) we obtain $J_0'(x) = J_{-1}(x)$. Thus $J_0'(x) = J_{-1}(x) = -J_1(x).$

33. Letting

$$s = \frac{2}{\alpha}\sqrt{\frac{k}{m}}\,e^{-\alpha t/2},$$

we have

$$\frac{dx}{dt} = \frac{dx}{ds}\frac{ds}{dt} = \frac{dx}{dt}\left[\frac{2}{\alpha}\sqrt{\frac{k}{m}}\left(-\frac{\alpha}{2}\right)e^{-\alpha t/2}\right] = \frac{dx}{ds}\left(-\sqrt{\frac{k}{m}}\,e^{-\alpha t/2}\right)$$

and

$$\frac{d^2x}{dt^2} = \frac{d}{dt}\left(\frac{dx}{dt}\right) = \frac{dx}{ds}\left(\frac{\alpha}{2}\sqrt{\frac{k}{m}}\,e^{-\alpha t/2}\right) + \frac{d}{dt}\left(\frac{dx}{ds}\right)\left(-\sqrt{\frac{k}{m}}\,e^{-\alpha t/2}\right)$$

$$= \frac{dx}{ds}\left(\frac{\alpha}{2}\sqrt{\frac{k}{m}}\,e^{-\alpha t/2}\right) + \frac{d^2x}{ds^2}\frac{ds}{dt}\left(-\sqrt{\frac{k}{m}}\,e^{-\alpha t/2}\right)$$

$$= \frac{dx}{ds}\left(\frac{\alpha}{2}\sqrt{\frac{k}{m}}\,e^{-\alpha t/2}\right) + \frac{d^2x}{ds^2}\left(\frac{k}{m}\,e^{-\alpha t}\right).$$

Then

$$m\frac{d^2x}{dt^2} + ke^{-\alpha t}x = ke^{-\alpha t}\frac{d^2x}{ds^2} + \frac{m\alpha}{2}\sqrt{\frac{k}{m}}\,e^{-\alpha t/2}\frac{dx}{ds} + ke^{-\alpha t}x = 0.$$

Multiplying by $2^2/\alpha^2 m$ we have

$$\frac{2^2}{\alpha^2}\frac{k}{m}e^{-\alpha t}\frac{d^2x}{ds^2} + \frac{2}{\alpha}\sqrt{\frac{k}{m}}\,e^{-\alpha t/2}\frac{dx}{ds} + \frac{2^2}{\alpha^2}\frac{k}{m}e^{-\alpha t}x = 0$$

or, since $s = (2/\alpha)\sqrt{k/m}\,e^{-\alpha t/2}$,

$$s^2\frac{d^2x}{ds^2} + s\frac{dx}{ds} + s^2x = 0.$$

36. The general solution of the differential equation is

$$y(x) = c_1 J_0(\alpha x) + c_2 Y_0(\alpha x).$$

In order to satisfy the conditions that $\lim_{x\to 0^+} y(x)$ and $\lim_{x\to 0^+} y'(x)$ are finite we are forced to define $c_2 = 0$. Thus, $y(x) = c_1 J_0(\alpha x)$. The second boundary condition, $y(2) = 0$, implies $c_1 = 0$ or $J_0(2\alpha) = 0$. In order to have a nontrivial solution we require that $J_0(2\alpha) = 0$. From Table 6.1, the first three positive zeros of J_0 are found to be

$$2\alpha_1 = 2.4048, \quad 2\alpha_2 = 5.5201, \quad 2\alpha_3 = 8.6537$$

and so $\alpha_1 = 1.2024$, $\alpha_2 = 2.7601$, $\alpha_3 = 4.3269$. The eigenfunctions corresponding to the eigenvalues $\lambda_1 = \alpha_1^2$, $\lambda_2 = \alpha_2^2$, $\lambda_3 = \alpha_3^2$ are $J_0(1.2024x)$, $J_0(2.7601x)$, and $J_0(4.3269x)$.

Legendre's Equation

45. The recurrence relation can be written

$$P_{k+1}(x) = \frac{2k+1}{k+1}xP_k(x) - \frac{k}{k+1}P_{k-1}(x), \qquad k = 2, 3, 4, \ldots .$$

$k = 1:\quad P_2(x) = \frac{3}{2}x^2 - \frac{1}{2}$

$k = 2:\quad P_3(x) = \frac{5}{3}x\left(\frac{3}{2}x^2 - \frac{1}{2}\right) - \frac{2}{3}x = \frac{5}{2}x^3 - \frac{3}{2}x$

$k = 3:\quad P_4(x) = \frac{7}{4}x\left(\frac{5}{2}x^3 - \frac{3}{2}x\right) - \frac{3}{4}\left(\frac{3}{2}x^2 - \frac{1}{2}\right) = \frac{35}{8}x^4 - \frac{30}{8}x^2 + \frac{3}{8}$

$k = 4:\quad P_5(x) = \frac{9}{5}x\left(\frac{35}{8}x^4 - \frac{30}{8}x^2 + \frac{3}{8}\right) - \frac{4}{5}\left(\frac{5}{2}x^3 - \frac{3}{2}x\right) = \frac{63}{8}x^5 - \frac{35}{4}x^3 + \frac{15}{8}x$

$$k = 5: \quad P_6(x) = \frac{11}{6}x\left(\frac{63}{8}x^5 - \frac{35}{4}x^3 + \frac{15}{8}x\right) - \frac{5}{6}\left(\frac{35}{8}x^4 - \frac{30}{8}x^2 + \frac{3}{8}\right)$$

$$= \frac{231}{16}x^6 - \frac{315}{16}x^4 + \frac{105}{16}x^2 - \frac{5}{16}$$

$$k = 6: \quad P_7(x) = \frac{13}{7}x\left(\frac{231}{16}x^6 - \frac{315}{16}x^4 + \frac{105}{16}x^2 - \frac{5}{16}\right) - \frac{6}{7}\left(\frac{63}{8}x^5 - \frac{35}{4}x^3 + \frac{15}{8}x\right)$$

$$= \frac{429}{16}x^7 - \frac{693}{16}x^5 + \frac{315}{16}x^3 - \frac{35}{16}x$$

Miscellaneous Differential Equations

51. Letting $y = \sum_{n=0}^{\infty} c_n x^n$ we have

$$y' = \sum_{n=1}^{\infty} nc_n x^{n-1} \qquad \text{and} \qquad y'' = \sum_{n=2}^{\infty} n(n+1)c_n x^{n-2}.$$

Then, with appropriate substitutions, we have

$$y'' - 2xy' + 2\alpha y = \sum_{k=0}^{\infty} \left[(k+2)(k+1)c_{k+2} + (2\alpha - 2k)c_k\right] x^k = 0.$$

This leads to the recurrence relation

$$c_{k+2} = -\frac{2\alpha - 2k}{(k+2)(k+1)} c_k, \quad \text{for } k = 0, 1, 2, 3, \ldots.$$

Thus

$$c_2 = -\frac{2\alpha}{2!} c_0$$

$$c_3 = -\frac{2\alpha - 2}{3!} c_1$$

$$c_4 = -\frac{2\alpha - 4}{4 \cdot 3} c_2 = \frac{2^2\alpha(\alpha - 2)}{4!} c_0$$

$$c_5 = -\frac{2\alpha - 6}{5 \cdot 4} c_3 = \frac{2^2(\alpha - 1)(\alpha - 3)}{5!} c_1$$

$$\vdots$$

Then two solutions are

$$y_1(x) = 1 - \frac{2\alpha}{2!}x^2 + \frac{2^2\alpha(\alpha - 2)}{4!}x^4 - \cdots = 1 + \sum_{k=1}^{\infty} \frac{(-1)^k 2^k \alpha(\alpha - 2) \cdots (\alpha - 2k + 2)}{(2k)!} x^{2k}$$

and

$$y_2(x) = x - \frac{2\alpha - 2}{3!}x^3 + \frac{2^2(\alpha - 1)(\alpha - 3)}{5!}x^5 - \cdots$$

$$= 1 + \sum_{k=1}^{\infty} \frac{(-1)^k 2^k (\alpha - 1)(\alpha - 3) \cdots (\alpha - 2k + 1)}{(2k + 1)!} x^{2k+1},$$

and the general solution is $y(x) = c_0 y_1(x) + c_1 y_2(x)$.

54. With $R(x) = (\alpha x)^{-1/2} Z(z)$ the product rule gives:

and
$$R' = \alpha^{-1/2}\left(x^{-1/2}Z' - \frac{1}{2}x^{-3/2}Z\right)$$

$$R'' = \alpha^{-1/2}\left(x^{-1/2}Z'' - \frac{1}{2}x^{-3/2}Z' - \frac{1}{2}x^{-3/2}Z' + \frac{3}{4}x^{-5/2}Z\right)$$

$$= \alpha^{-1/2}\left(x^{-1/2}Z'' - x^{-3/2}Z' + \frac{3}{4}x^{-5/2}Z\right).$$

The differential equation then becomes

$$\alpha^{-1/2}x^2\left(x^{-1/2}Z'' - x^{-3/2}Z' + \frac{3}{4}x^{-5/2}Z\right) + \alpha^{-1/2}2x\left(x^{-1/2}Z' - \frac{1}{2}x^{-3/2}Z\right)$$

$$+ \left[\alpha^2 x^2 - n(n+1)\right]\alpha^{-1/2}x^{-1/2}Z(x) = 0$$

or

$$x^2 Z'' + xZ' + \left[\alpha^2 x^2 - \left(n^2 + n + \frac{1}{4}\right)\right]Z = 0.$$

This is equivalent to

$$x^2 Z'' + xZ' + \left[\alpha^2 x^2 - \left(n + \frac{1}{2}\right)^2\right]Z = 0,$$

which is the parametric Bessel equation, so

$$Z(x) = C_1 J_{n+1/2}(\alpha x) + C_2 Y_{n+1/2}(\alpha x),$$

and

$$R(x) = \alpha^{-1/2}x^{1/2}Z(x) = \alpha^{-1/2}x^{-1/2}\left[C_1 J_{n+1/2}(\alpha x) + C_2 Y_{n+1/2}(\alpha x)\right].$$

Renaming C_1 and C_2 this becomes

$$R(x) = \left(c_1\sqrt{\frac{\pi}{2}}\right)\frac{J_{n+1/2}(\alpha x)}{\sqrt{\alpha x}} + \left(c_2\sqrt{\frac{\pi}{2}}\right)\frac{Y_{n+1/2}(\alpha x)}{\sqrt{\alpha x}}$$

$$= c_1\sqrt{\frac{\pi}{2\alpha x}}J_{n+1/2}(\alpha x) + c_2\sqrt{\frac{\pi}{2\alpha x}}Y_{n+1/2}(\alpha x)$$

$$= c_1 j_n(\alpha x) + c_2 y_n(\alpha x),$$

where $j_n(\alpha x)$ and $y_n(\alpha x)$ are the spherical Bessel functions of the first and second kind defined in the text.

6.R Chapter 6 in Review

3. $x = -1$ is the nearest singular point to the ordinary point $x = 0$. Theorem 6.2.1 guarantees the existence of two power series solutions $y = \sum_{n=1}^{\infty} c_n x^n$ of the differential equation that converge at least for $-1 < x < 1$. Since $-\frac{1}{2} \leq x \leq \frac{1}{2}$ is properly contained in $-1 < x < 1$, both power series must converge for all points contained in $-\frac{1}{2} \leq x \leq \frac{1}{2}$.

6. We have

$$f(x) = \frac{\sin x}{\cos x} = \frac{x - \dfrac{x^3}{6} + \dfrac{x^5}{120} - \cdots}{1 - \dfrac{x^2}{2} + \dfrac{x^4}{24} - \cdots} = x + \frac{x^3}{3} + \frac{2x^5}{15} + \cdots.$$

9. Substituting $y = \sum_{n=0}^{\infty} c_n x^{n+r}$ into the differential equation we obtain

$$2xy'' + y' + y = (2r^2 - r)\,c_0 x^{r-1} + \sum_{k=1}^{\infty} [2(k+r)(k+r-1)c_k + (k+r)c_k + c_{k-1}]x^{k+r-1} = 0$$

which implies

$$2r^2 - r = r(2r - 1) = 0$$

and

$$(k+r)(2k + 2r - 1)c_k + c_{k-1} = 0.$$

The indicial roots are $r = 0$ and $r = 1/2$. For $r = 0$ the recurrence relation is

$$c_k = -\frac{c_{k-1}}{k(2k-1)}, \quad k = 1, 2, 3, \ldots,$$

so

$$c_1 = -c_0, \qquad c_2 = \frac{1}{6}c_0, \qquad c_3 = -\frac{1}{90}c_0.$$

For $r = 1/2$ the recurrence relation is

$$c_k = -\frac{c_{k-1}}{k(2k+1)}, \quad k = 1, 2, 3, \ldots,$$

so

$$c_1 = -\frac{1}{3}c_0, \qquad c_2 = \frac{1}{30}c_0, \qquad c_3 = -\frac{1}{630}c_0.$$

Two linearly independent solutions are

$$y_1 = 1 - x + \frac{1}{6}x^2 - \frac{1}{90}x^3 + \cdots$$

and

$$y_2 = x^{1/2}\left(1 - \frac{1}{3}x + \frac{1}{30}x^2 - \frac{1}{630}x^3 + \cdots\right).$$

12. Substituting $y = \sum_{n=0}^{\infty} c_n x^n$ into the differential equation we obtain

$$y'' - x^2 y' + xy = 2c_2 + (6c_3 + c_0)x + \sum_{k=1}^{\infty} [(k+3)(k+2)c_{k+3} - (k-1)c_k]x^{k+1} = 0$$

which implies $c_2 = 0$, $c_3 = -c_0/6$, and

$$c_{k+3} = \frac{k-1}{(k+3)(k+2)} c_k, \quad k = 1, 2, 3, \ldots .$$

Choosing $c_0 = 1$ and $c_1 = 0$ we find

$$c_3 = -\frac{1}{6}$$

$$c_4 = c_7 = c_{10} = \cdots = 0$$

$$c_5 = c_8 = c_{11} = \cdots = 0$$

$$c_6 = -\frac{1}{90}$$

and so on. For $c_0 = 0$ and $c_1 = 1$ we obtain

$$c_3 = c_6 = c_9 = \cdots = 0$$

$$c_4 = c_7 = c_{10} = \cdots = 0$$

$$c_5 = c_8 = c_{11} = \cdots = 0$$

and so on. Thus, two solutions are

$$y_1 = 1 - \frac{1}{6}x^3 - \frac{1}{90}x^6 - \cdots \quad \text{and} \quad y_2 = x.$$

15. Substituting $y = \sum_{n=0}^{\infty} c_n x^n$ into the differential equation we have

$$y'' + xy' + 2y = \underbrace{\sum_{n=2}^{\infty} n(n-1)c_n x^{n-2}}_{k=n-2} + \underbrace{\sum_{n=1}^{\infty} n c_n x^n}_{k=n} + 2\underbrace{\sum_{n=0}^{\infty} c_n x^n}_{k=n}$$

$$= \sum_{k=0}^{\infty} (k+2)(k+1)c_{k+2}x^k + \sum_{k=1}^{\infty} k c_k x^k + 2\sum_{k=0}^{\infty} c_k x^k$$

$$= 2c_2 + 2c_0 + \sum_{k=1}^{\infty} [(k+2)(k+1)c_{k+2} + (k+2)c_k]x^k = 0$$

Thus

$$2c_2 + 2c_0 = 0$$

$$(k+2)(k+1)c_{k+2} + (k+2)c_k = 0$$

and

$$c_2 = -c_0$$

$$c_{k+2} = -\frac{1}{k+1} c_k, \quad k = 1, 2, 3, \ldots .$$

Choosing $c_0 = 1$ and $c_1 = 0$ we find

$$c_2 = -1$$

$$c_3 = c_5 = c_7 = \cdots = 0$$

$$c_4 = \frac{1}{3}$$

$$c_6 = -\frac{1}{15}$$

and so on. For $c_0 = 0$ and $c_1 = 1$ we obtain

$$c_2 = c_4 = c_6 = \cdots = 0$$

$$c_3 = -\frac{1}{2}$$

$$c_5 = \frac{1}{8}$$

$$c_7 = -\frac{1}{48}$$

and so on. Thus, the general solution is

$$y = C_0 \left(1 - x^2 + \frac{1}{3}x^4 - \frac{1}{15}x^6 + \cdots \right) + C_1 \left(x - \frac{1}{2}x^3 + \frac{1}{8}x^5 - \frac{1}{48}x^7 + \cdots \right)$$

and

$$y' = C_0 \left(-2x + \frac{4}{3}x^3 - \frac{2}{5}x^5 + \cdots \right) + C_1 \left(1 - \frac{3}{2}x^2 + \frac{5}{8}x^4 - \frac{7}{48}x^6 + \cdots \right).$$

Setting $y(0) = 3$ and $y'(0) = -2$ we find $c_0 = 3$ and $c_1 = -2$. Therefore, the solution of the initial-value problem is

$$y = 3 - 2x - 3x^2 + x^3 + x^4 - \frac{1}{4}x^5 - \frac{1}{5}x^6 + \frac{1}{24}x^7 + \cdots .$$

18. While we can find two solutions of the form

$$y_1 = c_0[1 + \cdots] \quad \text{and} \quad y_2 = c_1[x + \cdots],$$

the initial conditions at $x = 1$ give solutions for c_0 and c_1 in terms of infinite series. Letting $t = x - 1$ the initial-value problem becomes

$$\frac{d^2y}{dt^2} + (t+1)\frac{dy}{dt} + y = 0, \qquad y(0) = -6, \ y'(0) = 3.$$

Substituting $y = \sum_{n=0}^{\infty} c_n t^n$ into the differential equation we have

$$\frac{d^2y}{dt^2} + (t+1)\frac{dy}{dt} + y = \underbrace{\sum_{n=2}^{\infty} n(n-1)c_n t^{n-2}}_{k=n-2} + \underbrace{\sum_{n=1}^{\infty} nc_n t^n}_{k=n} + \underbrace{\sum_{n=1}^{\infty} nc_n t^{n-1}}_{k=n-1} + \underbrace{\sum_{n=0}^{\infty} c_n t^n}_{k=n}$$

$$= \sum_{k=0}^{\infty}(k+2)(k+1)c_{k+2}t^k + \sum_{k=1}^{\infty}kc_kt^k + \sum_{k=0}^{\infty}(k+1)c_{k+1}t^k + \sum_{k=0}^{\infty}c_kt^k$$

$$= 2c_2 + c_1 + c_0 + \sum_{k=1}^{\infty}[(k+2)(k+1)c_{k+2} + (k+1)c_{k+1} + (k+1)c_k]t^k = 0.$$

Thus

$$2c_2 + c_1 + c_0 = 0$$

$$(k+2)(k+1)c_{k+2} + (k+1)c_{k+1} + (k+1)c_k = 0$$

and

$$c_2 = -\frac{c_1 + c_0}{2}$$

$$c_{k+2} = -\frac{c_{k+1} + c_k}{k+2}, \quad k = 1, 2, 3, \dots .$$

Choosing $c_0 = 1$ and $c_1 = 0$ we find

$$c_2 = -\frac{1}{2}, \quad c_3 = \frac{1}{6}, \quad c_4 = \frac{1}{12},$$

and so on. For $c_0 = 0$ and $c_1 = 1$ we find

$$c_2 = -\frac{1}{2}, \quad c_3 = -\frac{1}{6}, \quad c_4 = \frac{1}{6},$$

and so on. Thus, the general solution is

$$y = c_0\left[1 - \frac{1}{2}t^2 + \frac{1}{6}t^3 + \frac{1}{12}t^4 + \cdots\right] + c_1\left[t - \frac{1}{2}t^2 - \frac{1}{6}t^3 + \frac{1}{6}t^4 + \cdots\right].$$

The initial conditions then imply $c_0 = -6$ and $c_1 = 3$. Thus the solution of the initial-value problem is

$$y = -6\left[1 - \frac{1}{2}(x-1)^2 + \frac{1}{6}(x-1)^3 + \frac{1}{12}(x-1)^4 + \cdots\right]$$

$$+ 3\left[(x-1) - \frac{1}{2}(x-1)^2 - \frac{1}{6}(x-1)^3 + \frac{1}{6}(x-1)^4 + \cdots\right].$$

21. Substituting $y = \sum_{n=0}^{\infty} c_n x^n$ into the differential equation we have

$$y'' + x^2y' + 2xy = \underbrace{\sum_{n=2}^{\infty} n(n-1)c_n x^{n-2}}_{k=n-2} + \underbrace{\sum_{n=1}^{\infty} nc_n x^{n+1}}_{k=n+1} + 2\underbrace{\sum_{n=0}^{\infty} c_n x^{n+1}}_{k=n+1}$$

$$= \sum_{k=0}^{\infty}(k+2)(k+1)c_{k+2}x^k + \sum_{k=2}^{\infty}(k-1)c_{k-1}x^k + 2\sum_{k=1}^{\infty}c_{k-1}x^k$$

$$= 2c_2 + (6c_3 + 2c_0)x + \sum_{k=2}^{\infty}[(k+2)(k+1)c_{k+2} + (k+1)c_{k-1}]x^k$$

$$= 5 - 2x + 10x^3.$$

Thus, equating coefficients of like powers of x gives

$$2c_2 = 5$$

$$6c_3 + 2c_0 = -2$$

$$12c_4 + 3c_1 = 0$$

$$20c_5 + 4c_2 = 10$$

$$(k+2)(k+1)c_{k+2} + (k+1)c_{k-1} = 0, \quad k = 4, 5, 6, \ldots,$$

and

$$c_2 = \frac{5}{2}$$

$$c_3 = -\frac{1}{3}c_0 - \frac{1}{3}$$

$$c_4 = -\frac{1}{4}c_1$$

$$c_5 = \frac{1}{2} - \frac{1}{5}c_2 = \frac{1}{2} - \frac{1}{5}\left(\frac{5}{2}\right) = 0$$

$$c_{k+2} = -\frac{1}{k+2}c_{k-1}.$$

Using the recurrence relation, we find

$$c_6 = -\frac{1}{6}c_3 = \frac{1}{3 \cdot 6}(c_0 + 1) = \frac{1}{3^2 \cdot 2!}c_0 + \frac{1}{3^2 \cdot 2!}$$

$$c_7 = -\frac{1}{7}c_4 = \frac{1}{4 \cdot 7}c_1$$

$$c_8 = c_{11} = c_{14} = \cdots = 0$$

$$c_9 = -\frac{1}{9}c_6 = -\frac{1}{3^3 \cdot 3!}c_0 - \frac{1}{3^3 \cdot 3!}$$

$$c_{10} = -\frac{1}{10}c_7 = -\frac{1}{4 \cdot 7 \cdot 10}c_1$$

$$c_{12} = -\frac{1}{12}c_9 = \frac{1}{3^4 \cdot 4!}c_0 + \frac{1}{3^4 \cdot 4!}$$

$$c_{13} = -\frac{1}{13}c_0 = \frac{1}{4 \cdot 7 \cdot 10 \cdot 13}c_1$$

and so on. Thus

$$y = c_0 \left[1 - \frac{1}{3}x^3 + \frac{1}{3^2 \cdot 2!}x^6 - \frac{1}{3^3 \cdot 3!}x^9 + \frac{1}{3^4 \cdot 4!}x^{12} - \cdots \right]$$

$$+ c_1 \left[x - \frac{1}{4}x^4 + \frac{1}{4 \cdot 7}x^7 - \frac{1}{4 \cdot 7 \cdot 10}x^{10} + \frac{1}{4 \cdot 7 \cdot 10 \cdot 13}x^{13} - \cdots \right]$$

$$+ \left[\frac{5}{2}x^2 - \frac{1}{3}x^3 + \frac{1}{3^2 \cdot 2!}x^6 - \frac{1}{3^3 \cdot 3!}x^9 + \frac{1}{3^4 \cdot 4!}x^{12} - \cdots \right].$$

24. **(a)** Using formula (5) of Section 4.2 in the text, we find that a second solution of $(1 - x^2)y'' - 2xy' = 0$ is

$$y_2(x) = 1 \cdot \int \frac{e^{\int 2x\,dx/(1-x^2)}}{1^2}\,dx = \int e^{-\ln(1-x^2)}\,dx$$

$$= \int \frac{dx}{1 - x^2} = \frac{1}{2}\ln\left(\frac{1+x}{1-x}\right),$$

where partial fractions was used to obtain the last integral.

(b) Using formula (5) of Section 4.2 in the text, we find that a second solution of $(1 - x^2)y'' - 2xy' + 2y = 0$ is

$$y_2(x) = x \cdot \int \frac{e^{\int 2x\,dx/(1-x^2)}}{x^2}\,dx = x \int \frac{e^{-\ln(1-x^2)}}{x^2}\,dx$$

$$= x \int \frac{dx}{x^2(1 - x^2)}\,dx = x \left[\frac{1}{2}\ln\left(\frac{1+x}{1-x}\right) - \frac{1}{x} \right]$$

$$= \frac{x}{2}\ln\left(\frac{1+x}{1-x}\right) - 1,$$

where partial fractions was used to obtain the last integral.

(c)

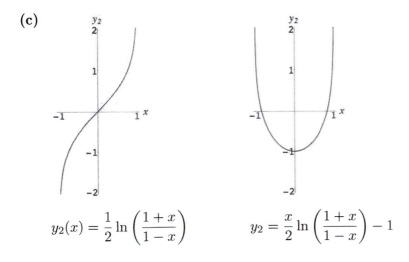

$$y_2(x) = \frac{1}{2}\ln\left(\frac{1+x}{1-x}\right) \qquad y_2 = \frac{x}{2}\ln\left(\frac{1+x}{1-x}\right) - 1$$

7 The Laplace Transform

The terminology and concepts listed below provide an outline of the main ideas encountered in this section. These can be useful when preparing for a quiz or test.

Terminology and Concepts

- integral transform
- definition of the Laplace transform
- linearity of the Laplace transform
- exponential order
- piecewise-continuous function
- transform of a piecewise-continuous function
- behavior of the Laplace transform $F(s)$ as $s \to \infty$

The basic skills listed below summarize the more mechanical types of problems encountered in the exercise set for this section.

Basic Skills

- find the Laplace transform of a function using the definition
- find the Laplace transform of a function using linearity and known transforms of some basic functions

Piecewise Continuity   A function f that is piecewise continuous on $[0, \infty)$ has only a finite number of *finite* discontinuities on any finite interval of $[0, \infty)$. Another name for a **finite discontinuity** is a **jump discontinuity**; this means that at any point of discontinuity t_1, the left- and right-hand limits, $\lim_{t \to t_1^-} f(t)$ and $\lim_{t \to t_1^+} f(t)$, respectively, exist but are different. Put another way, f cannot have any vertical asymptotes on the interval $[0, \infty)$.

Use of Computers   The Laplace transform function is built into version 5 of *Mathematica*. In *Maple* it is contained in the integral transforms package `inttrans`. The routines below define the Laplace transform of $f(t) = t^2 \cos t$ as a function $F(s)$.

Clear[f]   (*Mathematica*)
F[s_]:=Evaluate[LaplaceTransform[t^2 Cos[t], t, s]]
F[s]

with(inttrans);   (*Maple*)
F:=s->laplace(t^2*cos(t),t,s);
F(s);

3. $\mathscr{L}\{f(t)\} = \displaystyle\int_0^1 te^{-st}\,dt + \int_1^\infty e^{-st}\,dt = \left(-\frac{1}{s}te^{-st} - \frac{1}{s^2}e^{-st}\right)\Big|_0^1 - \frac{1}{s}e^{-st}\Big|_1^\infty$

$\qquad = \left(-\dfrac{1}{s}e^{-s} - \dfrac{1}{s^2}e^{-s}\right) - \left(0 - \dfrac{1}{s^2}\right) - \dfrac{1}{s}(0 - e^{-s}) = \dfrac{1}{s^2}(1 - e^{-s}), \quad s > 0$

6. $\mathscr{L}\{f(t)\} = \displaystyle\int_{\pi/2}^\infty (\cos t)e^{-st}\,dt = \left(-\frac{s}{s^2+1}e^{-st}\cos t + \frac{1}{s^2+1}e^{-st}\sin t\right)\Big|_{\pi/2}^\infty$

$\qquad = 0 - \left(0 + \dfrac{1}{s^2+1}e^{-\pi s/2}\right) = -\dfrac{1}{s^2+1}e^{-\pi s/2}, \quad s > 0$

9. The function is $f(t) = \begin{cases} 1 - t, & 0 < t < 1 \\ 0, & t > 1 \end{cases}$ so

$\mathscr{L}\{f(t)\} = \displaystyle\int_0^1 (1-t)e^{-st}\,dt + \int_1^\infty 0e^{-st}\,dt = \int_0^1 (1-t)e^{-st}\,dt = \left(-\frac{1}{s}(1-t)e^{-st} + \frac{1}{s^2}e^{-st}\right)\Big|_0^1$

$\qquad = \dfrac{1}{s^2}e^{-s} + \dfrac{1}{s} - \dfrac{1}{s^2}, \quad s > 0$

12. $\mathscr{L}\{f(t)\} = \displaystyle\int_0^\infty e^{-2t-5}e^{-st}\,dt = e^{-5}\int_0^\infty e^{-(s+2)t}\,dt = -\frac{e^{-5}}{s+2}e^{-(s+2)t}\Big|_0^\infty = \frac{e^{-5}}{s+2}, \quad s > -2$

15. $\mathscr{L}\{f(t)\} = \displaystyle\int_0^\infty e^{-t}(\sin t)e^{-st}\,dt = \int_0^\infty (\sin t)e^{-(s+1)t}\,dt$

$\qquad = \left(\dfrac{-(s+1)}{(s+1)^2+1}e^{-(s+1)t}\sin t - \dfrac{1}{(s+1)^2+1}e^{-(s+1)t}\cos t\right)\Big|_0^\infty$

$\qquad = \dfrac{1}{(s+1)^2+1} = \dfrac{1}{s^2+2s+2}, \quad s > -1$

18. $\mathscr{L}\{f(t)\} = \int_0^\infty t(\sin t)e^{-st}\,dt$

$$= \left[\left(-\frac{t}{s^2+1} - \frac{2s}{(s^2+1)^2}\right)(\cos t)e^{-st} - \left(\frac{st}{s^2+1} + \frac{s^2-1}{(s^2+1)^2}\right)(\sin t)e^{-st}\right]_0^\infty$$

$$= \frac{2s}{(s^2+1)^2}, \quad s > 0$$

21. $\mathscr{L}\{4t - 10\} = \dfrac{4}{s^2} - \dfrac{10}{s}$

24. $\mathscr{L}\{-4t^2 + 16t + 9\} = -4\dfrac{2}{s^3} + \dfrac{16}{s^2} + \dfrac{9}{s}$

27. $\mathscr{L}\{1 + e^{4t}\} = \dfrac{1}{s} + \dfrac{1}{s-4}$

30. $\mathscr{L}\{e^{2t} - 2 + e^{-2t}\} = \dfrac{1}{s-2} - \dfrac{2}{s} + \dfrac{1}{s+2}$

33. $\mathscr{L}\{\sinh kt\} = \dfrac{1}{2}\mathscr{L}\{e^{kt} - e^{-kt}\} = \dfrac{1}{2}\left[\dfrac{1}{s-k} - \dfrac{1}{s+k}\right] = \dfrac{k}{s^2 - k^2}$

36. $\mathscr{L}\{e^{-t}\cosh t\} = \mathscr{L}\left\{e^{-t}\dfrac{e^t + e^{-t}}{2}\right\} = \mathscr{L}\left\{\dfrac{1}{2} + \dfrac{1}{2}e^{-2t}\right\} = \dfrac{1}{2s} + \dfrac{1}{2(s+2)}$

39. From the addition formula for the sine function, $\sin(4t + 5) = (\sin 4t)(\cos 5) + (\cos 4t)(\sin 5)$ so

$$\mathscr{L}\{\sin(4t + 5)\} = (\cos 5)\mathscr{L}\{\sin 4t\} + (\sin 5)\mathscr{L}\{\cos 4t\}$$

$$= (\cos 5)\frac{4}{s^2 + 16} + (\sin 5)\frac{s}{s^2 + 16}$$

$$= \frac{4\cos 5 + (\sin 5)s}{s^2 + 16}.$$

42. Let $u = st$ so that $du = s\,dt$. Then

$$\mathscr{L}\{t^\alpha\} = \int_0^\infty e^{-st}t^\alpha\,dt = \int_0^\infty e^{-u}\left(\frac{u}{s}\right)^\alpha\frac{1}{s}\,du = \frac{1}{s^{\alpha+1}}\Gamma(\alpha + 1), \quad \alpha > -1.$$

45. $\mathscr{L}\{t^{3/2}\} = \dfrac{\Gamma(5/2)}{s^{5/2}} = \dfrac{3\sqrt{\pi}}{4s^{5/2}}$

7.2 | Inverse Transforms and Transforms of Derivatives

The terminology and concepts listed below provide an outline of the main ideas encountered in this section. These can be useful when preparing for a quiz or test.

Terminology and Concepts

- definition of the inverse Laplace transform
- linearity of the inverse Laplace transform
- Laplace transform of the derivative of a function
- use of the Laplace transform to solve an IVP

The basic skills listed below summarize the more mechanical types of problems encountered in the exercise set for this section.

Basic Skills

- find the inverse Laplace transform of a function (frequently using partial fractions)
- solve a linear IVP using the Laplace transform

Partial Fractions Again   The key to success in working with the Laplace transform is the ability to do algebra quickly and accurately. Partial fraction decomposition is used almost every time you have to solve some sort of equation. In this manual, check out the review of partial fractions given in Section 2.2.

Use of Computers   As can be seen in this section, partial fractions are used extensively in the computation of inverse Laplace transforms. This process can be very computationally intensive, and as such, is ideally suited to a CAS. The examples below show how to find the partial fraction decomposition of $(s^2 + 1)/(s^3 + 4s)$.

> Apart[(s^2+1)/(s^3+4s)]   (*Mathematica*)
>
> convert((s^2+1)/(s^3+4*s),parfrac,s);   (*Maple*)

The routines below define the inverse Laplace transform of $F(s) = (2s^3 - 6s)/(1 + s^2)^3$ as a function $f(t)$.

> Clear[f]   (*Mathematica*)
>
> f[t_]:=Evaluate[InverseLaplaceTransform[(2s^3 - 6s)/(s^2 + 1)^3, s, t]]
>
> f[t]

```
with(inttrans);                              (Maple)
f:=t->invlaplace((2*s^3-6*s)/(s^2+1)^3,s,t);
f(t);
```

3. $\mathcal{L}^{-1}\left\{\dfrac{1}{s^2} - \dfrac{48}{s^5}\right\} = \mathcal{L}^{-1}\left\{\dfrac{1}{s^2} - \dfrac{48}{24}\cdot\dfrac{4!}{s^5}\right\} = t - 2t^4$

6. $\mathcal{L}^{-1}\left\{\dfrac{(s+2)^2}{s^3}\right\} = \mathcal{L}^{-1}\left\{\dfrac{1}{s} + 4\cdot\dfrac{1}{s^2} + 2\cdot\dfrac{2}{s^3}\right\} = 1 + 4t + 2t^2$

9. $\mathcal{L}^{-1}\left\{\dfrac{1}{4s+1}\right\} = \dfrac{1}{4}\mathcal{L}^{-1}\left\{\dfrac{1}{s+1/4}\right\} = \dfrac{1}{4}e^{-t/4}$

12. $\mathcal{L}^{-1}\left\{\dfrac{10s}{s^2+16}\right\} = 10\cos 4t$

15. $\mathcal{L}^{-1}\left\{\dfrac{2s-6}{s^2+9}\right\} = \mathcal{L}^{-1}\left\{2\cdot\dfrac{s}{s^2+9} - 2\cdot\dfrac{3}{s^2+9}\right\} = 2\cos 3t - 2\sin 3t$

18. $\mathcal{L}^{-1}\left\{\dfrac{s+1}{s^2-4s}\right\} = \mathcal{L}^{-1}\left\{-\dfrac{1}{4}\cdot\dfrac{1}{s} + \dfrac{5}{4}\cdot\dfrac{1}{s-4}\right\} = -\dfrac{1}{4} + \dfrac{5}{4}e^{4t}$

21. $\mathcal{L}^{-1}\left\{\dfrac{0.9s}{(s-0.1)(s+0.2)}\right\} = \mathcal{L}^{-1}\left\{(0.3)\cdot\dfrac{1}{s-0.1} + (0.6)\cdot\dfrac{1}{s+0.2}\right\} = 0.3e^{0.1t} + 0.6e^{-0.2t}$

24. $\mathcal{L}^{-1}\left\{\dfrac{s^2+1}{s(s-1)(s+1)(s-2)}\right\} = \mathcal{L}^{-1}\left\{\dfrac{1}{2}\cdot\dfrac{1}{s} - \dfrac{1}{s-1} - \dfrac{1}{3}\cdot\dfrac{1}{s+1} + \dfrac{5}{6}\cdot\dfrac{1}{s-2}\right\}$

$$= \dfrac{1}{2} - e^t - \dfrac{1}{3}e^{-t} + \dfrac{5}{6}e^{2t}$$

27. $\mathcal{L}^{-1}\left\{\dfrac{2s-4}{(s^2+s)(s^2+1)}\right\} = \mathcal{L}^{-1}\left\{\dfrac{2s-4}{s(s+1)(s^2+1)}\right\} = \mathcal{L}^{-1}\left\{-\dfrac{4}{s} + \dfrac{3}{s+1} + \dfrac{s}{s^2+1} + \dfrac{3}{s^2+1}\right\}$

$$= -4 + 3e^{-t} + \cos t + 3\sin t$$

30. $\mathcal{L}^{-1}\left\{\dfrac{6s+3}{(s^2+1)(s^2+4)}\right\} = \mathcal{L}^{-1}\left\{2\cdot\dfrac{s}{s^2+1} + \dfrac{1}{s^2+1} - 2\cdot\dfrac{s}{s^2+4} - \dfrac{1}{2}\cdot\dfrac{2}{s^2+4}\right\}$

$$= 2\cos t + \sin t - 2\cos 2t - \dfrac{1}{2}\sin 2t$$

33. The Laplace transform of the initial-value problem is

$$s\mathcal{L}\{y\} - y(0) + 6\mathcal{L}\{y\} = \dfrac{1}{s-4}.$$

Solving for $\mathscr{L}\{y\}$ we obtain

$$\mathscr{L}\{y\} = \frac{1}{(s-4)(s+6)} + \frac{2}{s+6} = \frac{1}{10} \cdot \frac{1}{s-4} + \frac{19}{10} \cdot \frac{1}{s+6}.$$

Thus

$$y = \frac{1}{10}e^{4t} + \frac{19}{10}e^{-6t}.$$

36. The Laplace transform of the initial-value problem is

$$s^2 \mathscr{L}\{y\} - sy(0) - y'(0) - 4\left[s\,\mathscr{L}\{y\} - y(0)\right] = \frac{6}{s-3} - \frac{3}{s+1}.$$

Solving for $\mathscr{L}\{y\}$ we obtain

$$\mathscr{L}\{y\} = \frac{6}{(s-3)(s^2 - 4s)} - \frac{3}{(s+1)(s^2 - 4s)} + \frac{s-5}{s^2 - 4s}$$

$$= \frac{5}{2} \cdot \frac{1}{s} - \frac{2}{s-3} - \frac{3}{5} \cdot \frac{1}{s+1} + \frac{11}{10} \cdot \frac{1}{s-4}.$$

Thus

$$y = \frac{5}{2} - 2e^{3t} - \frac{3}{5}e^{-t} + \frac{11}{10}e^{4t}.$$

39. The Laplace transform of the initial-value problem is

$$2\left[s^3\mathscr{L}\{y\} - s^2 y(0) - sy'(0) - y''(0)\right] + 3[s^2\,\mathscr{L}\{y\} - sy(0) - y'(0)] - 3[s\,\mathscr{L}\{y\} - y(0)] - 2\,\mathscr{L}\{y\} = \frac{1}{s+1}.$$

Solving for $\mathscr{L}\{y\}$ we obtain

$$\mathscr{L}\{y\} = \frac{2s+3}{(s+1)(s-1)(2s+1)(s+2)} = \frac{1}{2}\frac{1}{s+1} + \frac{5}{18}\frac{1}{s-1} - \frac{8}{9}\frac{1}{s+1/2} + \frac{1}{9}\frac{1}{s+2}.$$

Thus

$$y = \frac{1}{2}e^{-t} + \frac{5}{18}e^{t} - \frac{8}{9}e^{-t/2} + \frac{1}{9}e^{-2t}.$$

42. The Laplace transform of the initial-value problem is

$$s^2\,\mathscr{L}\{y\} - s \cdot 1 - 3 - 2[s\,\mathscr{L}\{y\} - 1] + 5\,\mathscr{L}\{y\} = (s^2 - 2s + 5)\,\mathscr{L}\{y\} - s - 1 = 0.$$

Solving for $\mathscr{L}\{y\}$ we obtain

$$\mathscr{L}\{y\} = \frac{s+1}{s^2 - 2s + 5} = \frac{s-1+2}{(s-1)^2 + 2^2} = \frac{s-1}{(s-1)^2 + 2^2} + \frac{2}{(s-1)^2 + 2^2}.$$

Thus

$$y = e^{t}\cos 2t + e^{t}\sin 2t.$$

45. For $y'' - 4y' = 6e^{3t} - 3e^{-t}$ the transfer function is $W(s) = 1/(s^2 - 4s)$. The zero-input response is

$$y_0(t) = \mathscr{L}^{-1}\left\{\frac{s-5}{s^2-4s}\right\} = \mathscr{L}^{-1}\left\{\frac{5}{4}\cdot\frac{1}{s} - \frac{1}{4}\cdot\frac{1}{s-4}\right\} = \frac{5}{4} - \frac{1}{4}e^{4t},$$

and the zero-state response is

$$y_1(t) = \mathscr{L}^{-1}\left\{\frac{6}{(s-3)(s^2-4s)} - \frac{3}{(s+1)(s^2-4s)}\right\}$$

$$= \mathscr{L}^{-1}\left\{\frac{27}{20}\cdot\frac{1}{s-4} - \frac{2}{s-3} + \frac{5}{4}\cdot\frac{1}{s} - \frac{3}{5}\cdot\frac{1}{s+1}\right\}$$

$$= \frac{27}{20}e^{4t} - 2e^{3t} + \frac{5}{4} - \frac{3}{5}e^{-t}.$$

7.3 Operational Properties I

The terminology and concepts listed below provide an outline of the main ideas encountered in this section. These can be useful when preparing for a quiz or test.

Terminology and Concepts

- first translation (or shifting) theorem
- unit step function (Heaviside function)
- second translation (or shifting) theorem

The basic skills listed below summarize the more mechanical types of problems encountered in the exercise set for this section.

Basic Skills

- use the first translation theorem to find the Laplace transform of a function that is a multiple of e^{at}.

- use the first translation theorem to find the inverse Laplace transform of the shifted function $F(s-a)$

- use the Laplace transform in conjunction with the first translation theorem to solve an IVP

- express a piecewise-defined function in terms of unit step functions
- find the Laplace transform of a piecewise-defined function
- find the inverse Laplace transform of a function that is a multiple of e^{-as}
- use the Laplace transform to solve an IVP involving piecewise-defined functions

Use of Computers   See Section 2.3 in this manual for *Mathematica* and *Maple* syntax for the unit step function.

3. $\mathscr{L}\left\{t^3 e^{-2t}\right\} = \dfrac{3!}{(s+2)^4}$

6. $\mathscr{L}\left\{e^{2t}(t-1)^2\right\} = \mathscr{L}\left\{t^2 e^{2t} - 2te^{2t} + e^{2t}\right\} = \dfrac{2}{(s-2)^3} - \dfrac{2}{(s-2)^2} + \dfrac{1}{s-2}$

9. $\mathscr{L}\left\{(1 - e^t + 3e^{-4t})\cos 5t\right\} = \mathscr{L}\left\{\cos 5t - e^t \cos 5t + 3e^{-4t}\cos 5t\right\}$

$$= \dfrac{s}{s^2 + 25} - \dfrac{s-1}{(s-1)^2 + 25} + \dfrac{3(s+4)}{(s+4)^2 + 25}$$

12. $\mathscr{L}^{-1}\left\{\dfrac{1}{(s-1)^4}\right\} = \dfrac{1}{6}\mathscr{L}^{-1}\left\{\dfrac{3!}{(s-1)^4}\right\} = \dfrac{1}{6}t^3 e^t$

15. $\mathscr{L}^{-1}\left\{\dfrac{s}{s^2 + 4s + 5}\right\} = \mathscr{L}^{-1}\left\{\dfrac{s+2}{(s+2)^2 + 1^2} - 2\dfrac{1}{(s+2)^2 + 1^2}\right\} = e^{-2t}\cos t - 2e^{-2t}\sin t$

18. $\mathscr{L}^{-1}\left\{\dfrac{5s}{(s-2)^2}\right\} = \mathscr{L}^{-1}\left\{\dfrac{5(s-2) + 10}{(s-2)^2}\right\} = \mathscr{L}^{-1}\left\{\dfrac{5}{s-2} + \dfrac{10}{(s-2)^2}\right\} = 5e^{2t} + 10te^{2t}$

21. The Laplace transform of the differential equation is

$$s\mathscr{L}\{y\} - y(0) + 4\mathscr{L}\{y\} = \dfrac{1}{s+4}.$$

Solving for $\mathscr{L}\{y\}$ we obtain

$$\mathscr{L}\{y\} = \dfrac{1}{(s+4)^2} + \dfrac{2}{s+4}.$$

Thus

$$y = te^{-4t} + 2e^{-4t}.$$

24. The Laplace transform of the differential equation is

$$s^2 \mathcal{L}\{y\} - sy(0) - y'(0) - 4\left[s\mathcal{L}\{y\} - y(0)\right] + 4\mathcal{L}\{y\} = \frac{6}{(s-2)^4}.$$

Solving for $\mathcal{L}\{y\}$ we obtain $\mathcal{L}\{y\} = \dfrac{1}{20}\dfrac{5!}{(s-2)^6}$. Thus, $y = \dfrac{1}{20}t^5 e^{2t}$.

27. The Laplace transform of the differential equation is

$$s^2 \mathcal{L}\{y\} - sy(0) - y'(0) - 6\left[s\mathcal{L}\{y\} - y(0)\right] + 13\mathcal{L}\{y\} = 0.$$

Solving for $\mathcal{L}\{y\}$ we obtain

$$\mathcal{L}\{y\} = -\frac{3}{s^2 - 6s + 13} = -\frac{3}{2}\frac{2}{(s-3)^2 + 2^2}.$$

Thus

$$y = -\frac{3}{2}e^{3t}\sin 2t.$$

30. The Laplace transform of the differential equation is

$$s^2 \mathcal{L}\{y\} - sy(0) - y'(0) - 2\left[s\mathcal{L}\{y\} - y(0)\right] + 5\mathcal{L}\{y\} = \frac{1}{s} + \frac{1}{s^2}.$$

Solving for $\mathcal{L}\{y\}$ we obtain

$$\mathcal{L}\{y\} = \frac{4s^2 + s + 1}{s^2(s^2 - 2s + 5)} = \frac{7}{25}\frac{1}{s} + \frac{1}{5}\frac{1}{s^2} + \frac{-7s/25 + 109/25}{s^2 - 2s + 5}$$

$$= \frac{7}{25}\frac{1}{s} + \frac{1}{5}\frac{1}{s^2} - \frac{7}{25}\frac{s-1}{(s-1)^2 + 2^2} + \frac{51}{25}\frac{2}{(s-1)^2 + 2^2}.$$

Thus

$$y = \frac{7}{25} + \frac{1}{5}t - \frac{7}{25}e^t\cos 2t + \frac{51}{25}e^t\sin 2t.$$

33. Recall from Section 5.1 that $mx'' = -kx - \beta x'$. Now $m = W/g = 4/32 = \frac{1}{8}$ slug, and $4 = 2k$ so that $k = 2$ lb/ft. Thus, the differential equation is $x'' + 7x' + 16x = 0$. The initial conditions are $x(0) = -3/2$ and $x'(0) = 0$. The Laplace transform of the differential equation is

$$s^2\mathcal{L}\{x\} + \frac{3}{2}s + 7s\mathcal{L}\{x\} + \frac{21}{2} + 16\mathcal{L}\{x\} = 0.$$

Solving for $\mathcal{L}\{x\}$ we obtain

$$\mathcal{L}\{x\} = \frac{-3s/2 - 21/2}{s^2 + 7s + 16} = -\frac{3}{2}\frac{s + 7/2}{(s + 7/2)^2 + (\sqrt{15}/2)^2} - \frac{7\sqrt{15}}{10}\frac{\sqrt{15}/2}{(s + 7/2)^2 + (\sqrt{15}/2)^2}.$$

Thus

$$x = -\frac{3}{2}e^{-7t/2}\cos\frac{\sqrt{15}}{2}t - \frac{7\sqrt{15}}{10}e^{-7t/2}\sin\frac{\sqrt{15}}{2}t.$$

36. The differential equation is

$$R\frac{dq}{dt} + \frac{1}{C}q = E_0 e^{-kt}, \quad q(0) = 0.$$

The Laplace transform of this equation is

$$Rs\,\mathscr{L}\{q\} + \frac{1}{C}\mathscr{L}\{q\} = E_0\frac{1}{s+k}.$$

Solving for $\mathscr{L}\{q\}$ we obtain

$$\mathscr{L}\{q\} = \frac{E_0 C}{(s+k)(RCs+1)} = \frac{E_0/R}{(s+k)(s+1/RC)}.$$

When $1/RC \neq k$ we have by partial fractions

$$\mathscr{L}\{q\} = \frac{E_0}{R}\left(\frac{1/(1/RC-k)}{s+k} - \frac{1/(1/RC-k)}{s+1/RC}\right) = \frac{E_0}{R}\frac{1}{1/RC-k}\left(\frac{1}{s+k} - \frac{1}{s+1/RC}\right).$$

Thus

$$q(t) = \frac{E_0 C}{1-kRC}\left(e^{-kt} - e^{-t/RC}\right).$$

When $1/RC = k$ we have

$$\mathscr{L}\{q\} = \frac{E_0}{R}\frac{1}{(s+k)^2}.$$

Thus

$$q(t) = \frac{E_0}{R}te^{-kt} = \frac{E_0}{R}te^{-t/RC}.$$

39. $\mathscr{L}\{t\,\mathscr{U}(t-2)\} = \mathscr{L}\{(t-2)\,\mathscr{U}(t-2) + 2\,\mathscr{U}(t-2)\} = \dfrac{e^{-2s}}{s^2} + \dfrac{2e^{-2s}}{s}$

Alternatively, (16) of this section in the text could be used:

$$\mathscr{L}\{t\,\mathscr{U}(t-2)\} = e^{-2s}\mathscr{L}\{t+2\} = e^{-2s}\left(\frac{1}{s^2} + \frac{2}{s}\right).$$

42. $\mathscr{L}\left\{\sin t\,\mathscr{U}\left(t-\dfrac{\pi}{2}\right)\right\} = \mathscr{L}\left\{\cos\left(t-\dfrac{\pi}{2}\right)\mathscr{U}\left(t-\dfrac{\pi}{2}\right)\right\} = \dfrac{se^{-\pi s/2}}{s^2+1}$

Alternatively, (16) of this section in the text could be used:

$$\mathscr{L}\left\{\sin t\,\mathscr{U}\left(t-\frac{\pi}{2}\right)\right\} = e^{-\pi s/2}\mathscr{L}\left\{\sin\left(t+\frac{\pi}{2}\right)\right\} = e^{-\pi s/2}\mathscr{L}\{\cos t\} = e^{-\pi s/2}\frac{s}{s^2+1}.$$

45. $\mathcal{L}^{-1}\left\{\dfrac{e^{-\pi s}}{s^2+1}\right\} = \sin(t-\pi)\,\mathcal{U}(t-\pi) = -\sin t\,\mathcal{U}(t-\pi)$

48. $\mathcal{L}^{-1}\left\{\dfrac{e^{-2s}}{s^2(s-1)}\right\} = \mathcal{L}^{-1}\left\{-\dfrac{e^{-2s}}{s} - \dfrac{e^{-2s}}{s^2} + \dfrac{e^{-2s}}{s-1}\right\} = -\mathcal{U}(t-2) - (t-2)\,\mathcal{U}(t-2) + e^{t-2}\,\mathcal{U}(t-2)$

51. (f)   **54. (d)**

57. $\mathcal{L}\left\{t^2\,\mathcal{U}(t-1)\right\} = \mathcal{L}\left\{\left[(t-1)^2 + 2t - 1\right]\mathcal{U}(t-1)\right\} = \mathcal{L}\left\{\left[(t-1)^2 + 2(t-1) + 1\right]\mathcal{U}(t-1)\right\}$

$$= \left(\dfrac{2}{s^3} + \dfrac{2}{s^2} + \dfrac{1}{s}\right)e^{-s}$$

Alternatively, by (16) of this section in the text,

$$\mathcal{L}\left\{t^2\,\mathcal{U}(t-1)\right\} = e^{-s}\,\mathcal{L}\left\{t^2 + 2t + 1\right\} = e^{-s}\left(\dfrac{2}{s^3} + \dfrac{2}{s^2} + \dfrac{1}{s}\right).$$

60. $\mathcal{L}\left\{\sin t - \sin t\,\mathcal{U}(t-2\pi)\right\} = \mathcal{L}\left\{\sin t - \sin(t-2\pi)\,\mathcal{U}(t-2\pi)\right\} = \dfrac{1}{s^2+1} - \dfrac{e^{-2\pi s}}{s^2+1}$

63. The Laplace transform of the differential equation is

$$s\,\mathcal{L}\{y\} - y(0) + \mathcal{L}\{y\} = \dfrac{5}{s}e^{-s}.$$

Solving for $\mathcal{L}\{y\}$ we obtain

$$\mathcal{L}\{y\} = \dfrac{5e^{-s}}{s(s+1)} = 5e^{-s}\left[\dfrac{1}{s} - \dfrac{1}{s+1}\right].$$

Thus

$$y = 5\,\mathcal{U}(t-1) - 5e^{-(t-1)}\,\mathcal{U}(t-1).$$

66. The Laplace transform of the differential equation is

$$s^2\,\mathcal{L}\{y\} - sy(0) - y'(0) + 4\,\mathcal{L}\{y\} = \dfrac{1}{s} - \dfrac{e^{-s}}{s}.$$

Solving for $\mathcal{L}\{y\}$ we obtain

$$\mathcal{L}\{y\} = \dfrac{1-s}{s(s^2+4)} - e^{-s}\dfrac{1}{s(s^2+4)} = \dfrac{1}{4}\dfrac{1}{s} - \dfrac{1}{4}\dfrac{s}{s^2+4} - \dfrac{1}{2}\dfrac{2}{s^2+4} - e^{-s}\left[\dfrac{1}{4}\dfrac{1}{s} - \dfrac{1}{4}\dfrac{s}{s^2+4}\right].$$

Thus

$$y = \dfrac{1}{4} - \dfrac{1}{4}\cos 2t - \dfrac{1}{2}\sin 2t - \left[\dfrac{1}{4} - \dfrac{1}{4}\cos 2(t-1)\right]\mathcal{U}(t-1).$$

69. The Laplace transform of the differential equation is

$$s^2 \mathscr{L}\{y\} - sy(0) - y'(0) + \mathscr{L}\{y\} = \frac{e^{-\pi s}}{s} - \frac{e^{-2\pi s}}{s}.$$

Solving for $\mathscr{L}\{y\}$ we obtain

$$\mathscr{L}\{y\} = e^{-\pi s}\left[\frac{1}{s} - \frac{s}{s^2 + 1}\right] - e^{-2\pi s}\left[\frac{1}{s} - \frac{s}{s^2 + 1}\right] + \frac{1}{s^2 + 1}.$$

Thus

$$y = [1 - \cos(t - \pi)]\,\mathcal{U}(t - \pi) - [1 - \cos(t - 2\pi)]\,\mathcal{U}(t - 2\pi) + \sin t.$$

72. Recall from Section 5.1 that $mx'' = -kx + f(t)$. Now $m = W/g = 32/32 = 1$ slug, and $32 = 2k$ so that $k = 16$ lb/ft. Thus, the differential equation is $x'' + 16x = f(t)$. The initial conditions are $x(0) = 0$, $x'(0) = 0$. Also, since

$$f(t) = \begin{cases} \sin t, & 0 \le t < 2\pi \\ 0, & t \ge 2\pi \end{cases}$$

and $\sin t = \sin(t - 2\pi)$ we can write

$$f(t) = \sin t - \sin(t - 2\pi)\,\mathcal{U}(t - 2\pi).$$

The Laplace transform of the differential equation is

$$s^2 \mathscr{L}\{x\} + 16\mathscr{L}\{x\} = \frac{1}{s^2 + 1} - \frac{1}{s^2 + 1}e^{-2\pi s}.$$

Solving for $\mathscr{L}\{x\}$ we obtain

$$\mathscr{L}\{x\} = \frac{1}{(s^2 + 16)\,(s^2 + 1)} - \frac{1}{(s^2 + 16)\,(s^2 + 1)}e^{-2\pi s}$$

$$= \frac{-1/15}{s^2 + 16} + \frac{1/15}{s^2 + 1} - \left[\frac{-1/15}{s^2 + 16} + \frac{1/15}{s^2 + 1}\right]e^{-2\pi s}.$$

Thus

$$x(t) = -\frac{1}{60}\sin 4t + \frac{1}{15}\sin t + \frac{1}{60}\sin 4(t - 2\pi)\,\mathcal{U}(t - 2\pi) - \frac{1}{15}\sin(t - 2\pi)\,\mathcal{U}(t - 2\pi)$$

$$= \begin{cases} -\frac{1}{60}\sin 4t + \frac{1}{15}\sin t, & 0 \le t < 2\pi \\ 0, & t \ge 2\pi. \end{cases}$$

75. (a) The differential equation is

$$\frac{di}{dt} + 10i = \sin t + \cos\left(t - \frac{3\pi}{2}\right)\mathcal{U}\left(t - \frac{3\pi}{2}\right), \quad i(0) = 0.$$

The Laplace transform of this equation is

$$s\mathscr{L}\{i\} + 10\mathscr{L}\{i\} = \frac{1}{s^2 + 1} + \frac{se^{-3\pi s/2}}{s^2 + 1}.$$

Solving for $\mathscr{L}\{i\}$ we obtain

$$\mathscr{L}\{i\} = \frac{1}{(s^2+1)(s+10)} + \frac{s}{(s^2+1)(s+10)} e^{-3\pi s/2}$$

$$= \frac{1}{101}\left(\frac{1}{s+10} - \frac{s}{s^2+1} + \frac{10}{s^2+1}\right) + \frac{1}{101}\left(\frac{-10}{s+10} + \frac{10s}{s^2+1} + \frac{1}{s^2+1}\right)e^{-3\pi s/2}.$$

Thus

$$i(t) = \frac{1}{101}\left(e^{-10t} - \cos t + 10\sin t\right)$$

$$+ \frac{1}{101}\left[-10e^{-10(t-3\pi/2)} + 10\cos\left(t - \frac{3\pi}{2}\right) + \sin\left(t - \frac{3\pi}{2}\right)\right]\mathscr{U}\left(t - \frac{3\pi}{2}\right).$$

(b)

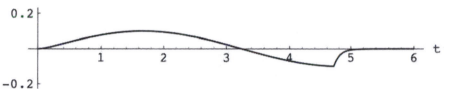

The maximum value of $i(t)$ is approximately 0.1 at $t = 1.7$, the minimum is approximately -0.1 at 4.7. [Using *Mathematica* we see that the maximum value is $i(t)$ is 0.0995037 at $t = 1.670465$, and the mininum value is $i(3\pi/2) \approx -0.0990099$ at $t = 3\pi/2$.]

78. The differential equation is

$$EI\frac{d^4y}{dx^4} = w_0[\mathscr{U}(x - L/3) - \mathscr{U}(x - 2L/3)].$$

Taking the Laplace transform of both sides and using $y(0) = y'(0) = 0$ we obtain

$$s^4\mathscr{L}\{y\} - sy''(0) - y'''(0) = \frac{w_0}{EI}\frac{1}{s}\left(e^{-Ls/3} - e^{-2Ls/3}\right).$$

Letting $y''(0) = c_1$ and $y'''(0) = c_2$ we have

$$\mathscr{L}\{y\} = \frac{c_1}{s^3} + \frac{c_2}{s^4} + \frac{w_0}{EI}\frac{1}{s^5}\left(e^{-Ls/3} - e^{-2Ls/3}\right)$$

so that

$$y(x) = \frac{1}{2}c_1x^2 + \frac{1}{6}c_2x^3 + \frac{1}{24}\frac{w_0}{EI}\left[\left(x - \frac{L}{3}\right)^4\mathscr{U}\left(x - \frac{L}{3}\right) - \left(x - \frac{2L}{3}\right)^4\mathscr{U}\left(x - \frac{2L}{3}\right)\right].$$

To find c_1 and c_2 we compute

$$y''(x) = c_1 + c_2x + \frac{1}{2}\frac{w_0}{EI}\left[\left(x - \frac{L}{3}\right)^2\mathscr{U}\left(x - \frac{L}{3}\right) - \left(x - \frac{2L}{3}\right)^2\mathscr{U}\left(x - \frac{2L}{3}\right)\right]$$

and

$$y'''(x) = c_2 + \frac{w_0}{EI}\left[\left(x - \frac{L}{3}\right)\mathscr{U}\left(x - \frac{L}{3}\right) - \left(x - \frac{2L}{3}\right)\mathscr{U}\left(x - \frac{2L}{3}\right)\right].$$

Then $y''(L) = y'''(L) = 0$ yields the system

$$c_1 + c_2 L + \frac{1}{2}\frac{w_0}{EI}\left[\left(\frac{2L}{3}\right)^2 - \left(\frac{L}{3}\right)^2\right] = c_1 + c_2 L + \frac{1}{6}\frac{w_0 L^2}{EI} = 0$$

$$c_2 + \frac{w_0}{EI}\left[\frac{2L}{3} - \frac{L}{3}\right] = c_2 + \frac{1}{3}\frac{w_0 L}{EI} = 0.$$

Solving for c_1 and c_2 we obtain $c_1 = \frac{1}{6}w_0 L^2/EI$ and $c_2 = -\frac{1}{3}w_0 L/EI$. Thus

$$y(x) = \frac{w_0}{EI}\left(\frac{1}{12}L^2 x^2 - \frac{1}{18}Lx^3 + \frac{1}{24}\left[\left(x - \frac{L}{3}\right)^4 \mathcal{U}\left(x - \frac{L}{3}\right) - \left(x - \frac{2L}{3}\right)^4 \mathcal{U}\left(x - \frac{2L}{3}\right)\right]\right).$$

81. **(a)** The temperature T of the cake inside the oven is modeled by

$$\frac{dT}{dt} = k(T - T_m)$$

where T_m is the ambient temperature of the oven. For $0 \le t \le 4$, we have

$$T_m = 70 + \frac{300 - 70}{4 - 0}t = 70 + 57.5t.$$

Hence for $t \ge 0$,

$$T_m = \begin{cases} 70 + 57.5t, & 0 \le t < 4 \\ 300, & t \ge 4. \end{cases}$$

In terms of the unit step function,

$$T_m = (70 + 57.5t)[1 - \mathcal{U}(t - 4)] + 300\,\mathcal{U}(t - 4) = 70 + 57.5t + (230 - 57.5t)\,\mathcal{U}(t - 4).$$

The initial-value problem is then

$$\frac{dT}{dt} = k[T - 70 - 57.5t - (230 - 57.5t)\,\mathcal{U}(t - 4)], \qquad T(0) = 70.$$

(b) Let $t(s) = \mathcal{L}\{T(t)\}$. Transforming the equation, using $230 - 57.5t = -57.5(t - 4)$ and Theorem 7.3.2, gives

$$st(s) - 70 = k\left(t(s) - \frac{70}{s} - \frac{57.5}{s^2} + \frac{57.5}{s^2}e^{-4s}\right)$$

or

$$t(s) = \frac{70}{s - k} - \frac{70k}{s(s - k)} - \frac{57.5k}{s^2(s - k)} + \frac{57.5k}{s^2(s - k)}e^{-4s}.$$

After using partial functions, the inverse transform is then

$$T(t) = 70 + 57.5\left(\frac{1}{k} + t - \frac{1}{k}e^{kt}\right) - 57.5\left(\frac{1}{k} + t - 4 - \frac{1}{k}e^{k(t-4)}\right)\mathcal{U}(t - 4).$$

Of course, the obvious question is: What is k? If the cake is supposed to bake for, say, 20 minutes, then $T(20) = 300$. That is,

$$300 = 70 + 57.5\left(\frac{1}{k} + 20 - \frac{1}{k}e^{20k}\right) - 57.5\left(\frac{1}{k} + 16 - \frac{1}{k}e^{16k}\right).$$

But this equation has no physically meaningful solution. This should be no surprise since the model predicts the asymptotic behavior $T(t) \to 300$ as t increases. Using $T(20) = 299$ instead, we find, with the help of a CAS, that $k \approx -0.3$.

7.4 | Operational Properties II

The terminology and concepts listed below provide an outline of the main ideas encountered in this section. These can be useful when preparing for a quiz or test.

Terminology and Concepts

- the relationship between $t^n f(t)$ and the derivative of the Laplace transform of $f(t)$
- convolution of two functions
- the Laplace transform of the convolution f*g of two functions
- the Laplace transform of the integral of a function
- Volterra integral equation
- integrodifferential differential equation
- the Laplace transform of a periodic function

The basic skills listed below summarize the more mechanical types of problems encountered in the exercise set for this section.

Basic Skills

- find the Laplace transform of the product $t^n f(t)$ in terms of the Laplace transform of $f(t)$
- find the Laplace transform of the convolution of two functions
- find the Laplace transform of the integral of a function
- solve an integral equation
- solve an integrodifferential equation
- find the Laplace transform of a periodic function

DEs With Variable Coefficients

The Laplace transform is not well-suited to solving linear differential equations with variable coefficients. Nevertheless, it can be used in *some* instances. If the DE has coefficients that are at most linear polynomials in t, then its transform is a linear first-order DE in the transformed function $Y(s)$. For example, let's find a solution of $ty'' + 2ty' + 2y = t$ subject to $y(0) = 0$. From Theorem 7.4.1 observe

$$\mathcal{L}\{ty''\} = -\frac{d}{ds}\mathcal{L}\{y''\} = -\frac{d}{ds}[s^2 Y(s) - sy(0) - y'(0)] = -s^2 Y'(s) - 2sY(s) + y(0) \qquad (1)$$

$$\mathcal{L}\{ty'\} = -\frac{d}{ds}\mathcal{L}\{y'\} = -\frac{d}{ds}[sY(s) - y(0)] = -sY'(s) - Y(s). \qquad (2)$$

Note here that we have used the product rule to differentiate $s^2Y(s)$ and $sY(s)$. Hence the transform of the second-order DE is the first-order DE,

$$-s^2Y'(s) - 2sY(s) - 2sY'(s) - 2Y(s) + 2Y(s) = \frac{1}{s^2}$$

or

$$\frac{dY}{ds} + \frac{2}{s+2}Y = -\frac{1}{s^3(s+2)},$$

where we have used $y(0) = 0$. Since this last equation is recognized as a linear first-order DE in Y, we find from Section 2.3 in the text that the integrating factor is $(s+2)^2$. Multiplying the last equation by this factor gives

$$\frac{d}{ds}[(s+2)^2Y] = -\frac{s+2}{s^3} = -\frac{1}{s^2} - \frac{2}{s^3}.$$

Integrating, we obtain

$$(s+2)^2Y = \frac{1}{s} + \frac{1}{s^2} + c = \frac{s+1}{s^2} + c.$$

Now solve for Y and expand into partial fractions:

$$Y = \frac{s+1}{s^2(s+2)^2} + \frac{c}{(s+2)^2} = \frac{1}{4}\frac{1}{s^2} - \frac{1}{4}\frac{1}{(s+2)^2} + \frac{c}{(s+2)^2}.$$

The inverse of the preceding expression is then

$$y(t) = \frac{1}{4}t - \frac{1}{4}te^{-2t} + cte^{-2t} \qquad \text{or} \qquad y(t) = \frac{1}{4}t + c_1te^{-2t},$$

where we have replaced the term $c - \frac{1}{4}$ by c_1.

7.4.1   DERIVATIVES OF A TRANSFORM

3. $\mathscr{L}\{t\cos 2t\} = -\dfrac{d}{ds}\left(\dfrac{s}{s^2+4}\right) = \dfrac{s^2-4}{(s^2+4)^2}$

6. $\mathscr{L}\{t^2\cos t\} = \dfrac{d^2}{ds^2}\left(\dfrac{s}{s^2+1}\right) = \dfrac{d}{ds}\left(\dfrac{1-s^2}{(s^2+1)^2}\right) = \dfrac{2s(s^2-3)}{(s^2+1)^3}$

9. The Laplace transform of the differential equation is

$$s\mathscr{L}\{y\} + \mathscr{L}\{y\} = \frac{2s}{(s^2+1)^2}.$$

Solving for $\mathscr{L}\{y\}$ we obtain

$$\mathscr{L}\{y\} = \frac{2s}{(s+1)(s^2+1)^2} = -\frac{1}{2}\frac{1}{s+1} - \frac{1}{2}\frac{1}{s^2+1} + \frac{1}{2}\frac{s}{s^2+1} + \frac{1}{(s^2+1)^2} + \frac{s}{(s^2+1)^2}.$$

Thus

$$y(t) = -\frac{1}{2}e^{-t} - \frac{1}{2}\sin t + \frac{1}{2}\cos t + \frac{1}{2}(\sin t - t\cos t) + \frac{1}{2}t\sin t$$

$$= -\frac{1}{2}e^{-t} + \frac{1}{2}\cos t - \frac{1}{2}t\cos t + \frac{1}{2}t\sin t.$$

12. The Laplace transform of the differential equation is

$$s^2\mathscr{L}\{y\} - sy(0) - y'(0) + \mathscr{L}\{y\} = \frac{1}{s^2 + 1}.$$

Solving for $\mathscr{L}\{y\}$ we obtain

$$\mathscr{L}\{y\} = \frac{s^3 - s^2 + s}{(s^2 + 1)^2} = \frac{s}{s^2 + 1} - \frac{1}{s^2 + 1} + \frac{1}{(s^2 + 1)^2}.$$

Thus

$$y = \cos t - \sin t + \left(\frac{1}{2}\sin t - \frac{1}{2}t\cos t\right) = \cos t - \frac{1}{2}\sin t - \frac{1}{2}t\cos t.$$

15.

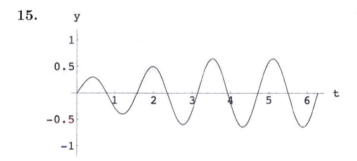

18. From Theorem 7.4.1 in the text

$$\mathscr{L}\{ty'\} = -\frac{d}{ds}\mathscr{L}\{y'\} = -\frac{d}{ds}[sY(s) - y(0)] = -s\frac{dY}{ds} - Y$$

so that the transform of the given second-order differential equation is the linear first-order differential equation in $Y(s)$:

$$Y' + \left(\frac{3}{s} - 2s\right)Y = -\frac{10}{s}.$$

Using the integrating factor $s^3 e^{-s^2}$, the last equation yields

$$Y(s) = \frac{5}{s^3} + \frac{c}{s^3}e^{s^2}.$$

But if $Y(s)$ is the Laplace transform of a piecewise-continuous function of exponential order, we must have, in view of Theorem 7.1.3, $\lim_{s\to\infty} Y(s) = 0$. In order to obtain this condition we require $c = 0$. Hence

$$y(t) = \mathscr{L}^{-1}\left\{\frac{5}{s^3}\right\} = \frac{5}{2}t^2.$$

7.4.2   TRANSFORMS OF INTEGRALS

21. $\mathscr{L}\{e^{-t} * e^{t}\cos t\} = \dfrac{s-1}{(s+1)\left[(s-1)^2+1\right]}$

24. $\mathscr{L}\left\{\displaystyle\int_0^t \cos\tau\, d\tau\right\} = \dfrac{1}{s}\mathscr{L}\{\cos t\} = \dfrac{s}{s(s^2+1)} = \dfrac{1}{s^2+1}$

27. $\mathscr{L}\left\{\displaystyle\int_0^t \tau e^{t-\tau}\, d\tau\right\} = \mathscr{L}\{t\}\,\mathscr{L}\{e^{t}\} = \dfrac{1}{s^2(s-1)}$

30. $\mathscr{L}\left\{t\displaystyle\int_0^t \tau e^{-\tau}\,d\tau\right\} = -\dfrac{d}{ds}\mathscr{L}\left\{\displaystyle\int_0^t \tau e^{-\tau}\,d\tau\right\} = -\dfrac{d}{ds}\left(\dfrac{1}{s}\dfrac{1}{(s+1)^2}\right) = \dfrac{3s+1}{s^2(s+1)^3}$

33. $\mathscr{L}^{-1}\left\{\dfrac{1}{s^3(s-1)}\right\} = \mathscr{L}^{-1}\left\{\dfrac{1/s^2(s-1)}{s}\right\} = \displaystyle\int_0^t (e^{\tau}-\tau-1)\,d\tau = e^{t} - \dfrac{1}{2}t^2 - t - 1$

36. The Laplace transform of the differential equation is

$$s^2\mathscr{L}\{y\} + \mathscr{L}\{y\} = \dfrac{1}{(s^2+1)} + \dfrac{2s}{(s^2+1)^2}.$$

Thus

$$\mathscr{L}\{y\} = \dfrac{1}{(s^2+1)^2} + \dfrac{2s}{(s^2+1)^3}$$

and, using Problem 35 with $k=1$,

$$y = \dfrac{1}{2}(\sin t - t\cos t) + \dfrac{1}{4}(t\sin t - t^2\cos t).$$

39. The Laplace transform of the given equation is

$$\mathscr{L}\{f\} = \mathscr{L}\{te^{t}\} + \mathscr{L}\{t\}\,\mathscr{L}\{f\}.$$

Solving for $\mathscr{L}\{f\}$ we obtain

$$\mathscr{L}\{f\} = \dfrac{s^2}{(s-1)^3(s+1)} = \dfrac{1}{8}\dfrac{1}{s-1} + \dfrac{3}{4}\dfrac{1}{(s-1)^2} + \dfrac{1}{4}\dfrac{2}{(s-1)^3} - \dfrac{1}{8}\dfrac{1}{s+1}.$$

Thus

$$f(t) = \dfrac{1}{8}e^{t} + \dfrac{3}{4}te^{t} + \dfrac{1}{4}t^2e^{t} - \dfrac{1}{8}e^{-t}$$

42. The Laplace transform of the given equation is

$$\mathscr{L}\{f\} = \mathscr{L}\{\cos t\} + \mathscr{L}\{e^{-t}\}\,\mathscr{L}\{f\}.$$

Solving for $\mathscr{L}\{f\}$ we obtain

$$\mathscr{L}\{f\} = \frac{s}{s^2+1} + \frac{1}{s^2+1}.$$

Thus

$$f(t) = \cos t + \sin t.$$

45. The Laplace transform of the given equation is

$$s\mathscr{L}\{y\} - y(0) = \mathscr{L}\{1\} - \mathscr{L}\{\sin t\} - \mathscr{L}\{1\}\,\mathscr{L}\{y\}.$$

Solving for $\mathscr{L}\{f\}$ we obtain

$$\mathscr{L}\{y\} = \frac{s^2-s+1}{(s^2+1)^2} = \frac{1}{s^2+1} - \frac{1}{2}\frac{2s}{(s^2+1)^2}.$$

Thus

$$y = \sin t - \frac{1}{2}\,t\sin t.$$

48. The differential equation is

$$0.005\frac{di}{dt} + i + \frac{1}{0.02}\int_0^t i(\tau)d\tau = 100\big[t - (t-1)\mathscr{U}(t-1)\big]$$

or

$$\frac{di}{dt} + 200i + 10{,}000\int_0^t i(\tau)d\tau = 20{,}000\big[t - (t-1)\mathscr{U}(t-1)\big],$$

where $i(0) = 0$. The Laplace transform of the differential

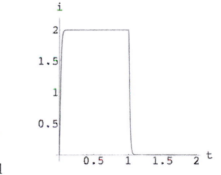

equation is

$$s\mathscr{L}\{i\} + 200\mathscr{L}\{i\} + \frac{10{,}000}{s}\mathscr{L}\{i\} = 20{,}000\left(\frac{1}{s^2} - \frac{1}{s^2}e^{-s}\right).$$

Solving for $\mathscr{L}\{i\}$ we obtain

$$\mathscr{L}\{i\} = \frac{20{,}000}{s(s+100)^2}(1 - e^{-s}) = \left[\frac{2}{s} - \frac{2}{s+100} - \frac{200}{(s+100)^2}\right](1 - e^{-s}).$$

Thus

$$i(t) = 2 - 2e^{-100t} - 200te^{-100t} - 2\mathscr{U}(t-1)) + 2e^{-100(t-1)}\mathscr{U}(t-1))$$
$$+ 200(t-1)e^{-100(t-1)}\mathscr{U}(t-1).$$

7.4.2 TRANSFORM OF A PERIODIC FUNCTION

51. Using integration by parts,

$$\mathscr{L}\{f(t)\} = \frac{1}{1-e^{-bs}} \int_0^b \frac{a}{b} t e^{-st} dt = \frac{a}{s}\left(\frac{1}{bs} - \frac{1}{e^{bs}-1}\right).$$

54. $\mathscr{L}\{f(t)\} = \dfrac{1}{1-e^{-2\pi s}} \displaystyle\int_0^\pi e^{-st} \sin t\, dt = \dfrac{1}{s^2+1} \cdot \dfrac{1}{1-e^{-\pi s}}$

57. The differential equation is $x'' + 2x' + 10x = 20f(t)$, where $f(t)$ is the meander function in Problem 49 with $a = \pi$. Using the initial conditions $x(0) = x'(0) = 0$ and taking the Laplace transform we obtain

$$(s^2 + 2s + 10)\mathscr{L}\{x(t)\} = \frac{20}{s}(1 - e^{-\pi s})\frac{1}{1+e^{-\pi s}}$$

$$= \frac{20}{s}(1 - e^{-\pi s})(1 - e^{-\pi s} + e^{-2\pi s} - e^{-3\pi s} + \cdots)$$

$$= \frac{20}{s}(1 - 2e^{-\pi s} + 2e^{-2\pi s} - 2e^{-3\pi s} + \cdots)$$

$$= \frac{20}{s} + \frac{40}{s}\sum_{n=1}^\infty (-1)^n e^{-n\pi s}.$$

Then

$$\mathscr{L}\{x(t)\} = \frac{20}{s(s^2 + 2s + 10)} + \frac{40}{s(s^2 + 2s + 10)}\sum_{n=1}^\infty (-1)^n e^{-n\pi s}$$

$$= \frac{2}{s} - \frac{2s+4}{s^2 + 2s + 10} + \sum_{n=1}^\infty (-1)^n \left[\frac{4}{s} - \frac{4s+8}{s^2 + 2s + 10}\right] e^{-n\pi s}$$

$$= \frac{2}{s} - \frac{2(s+1)+2}{(s+1)^2 + 9} + 4\sum_{n=1}^\infty (-1)^n \left[\frac{1}{s} - \frac{(s+1)+1}{(s+1)^2 + 9}\right] e^{-n\pi s}$$

and

$$x(t) = 2\left(1 - e^{-t}\cos 3t - \frac{1}{3}e^{-t}\sin 3t\right)$$

$$+ 4\sum_{n=1}^\infty (-1)^n \left[1 - e^{-(t-n\pi)}\cos 3(t - n\pi) - \frac{1}{3}e^{-(t-n\pi)}\sin 3(t - n\pi)\right]\mathscr{U}(t - n\pi).$$

The graph of $x(t)$ on the interval $[0, 2\pi)$ is shown below.

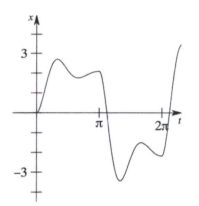

7.5 | The Dirac Delta Function

The terminology and concepts listed below provide an outline of the main ideas encountered in this section. These can be useful when preparing for a quiz or test.

Terminology and Concepts

- unit impulse function

- Dirac delta function

- Laplace transform of the Dirac delta function

The basic skills listed below summarize the more mechanical types of problems encountered in the exercise set for this section.

Basic Skills

- solve a linear second-order DE whose input function involves the Dirac delta function

3. The Laplace transform of the differential equation yields

$$\mathscr{L}\{y\} = \frac{1}{s^2 + 1}\left(1 + e^{-2\pi s}\right)$$

so that

$$y = \sin t + \sin t\, \mathscr{U}(t - 2\pi).$$

6. The Laplace transform of the differential equation yields

$$\mathscr{L}\{y\} = \frac{s}{s^2 + 1} + \frac{1}{s^2 + 1}(e^{-2\pi s} + e^{-4\pi s})$$

so that

$$y = \cos t + \sin t[\mathscr{U}(t - 2\pi) + \mathscr{U}(t - 4\pi)].$$

9. The Laplace transform of the differential equation yields

$$\mathscr{L}\{y\} = \frac{1}{(s + 2)^2 + 1}e^{-2\pi s}$$

so that

$$y = e^{-2(t-2\pi)}\sin t\, \mathscr{U}(t - 2\pi).$$

12. The Laplace transform of the differential equation yields

$$\mathscr{L}\{y\} = \frac{1}{(s - 1)^2(s - 6)} + \frac{e^{-2s} + e^{-4s}}{(s - 1)(s - 6)}$$

$$= -\frac{1}{25}\frac{1}{s - 1} - \frac{1}{5}\frac{1}{(s - 1)^2} + \frac{1}{25}\frac{1}{s - 6} + \left[-\frac{1}{5}\frac{1}{s - 1} + \frac{1}{5}\frac{1}{s - 6}\right](e^{-2s} + e^{-4s})$$

so that

$$y = -\frac{1}{25}e^t - \frac{1}{5}te^t + \frac{1}{25}e^{6t} + \left[-\frac{1}{5}e^{t-2} + \frac{1}{5}e^{6(t-2)}\right]\mathscr{U}(t - 2)$$

$$+ \left[-\frac{1}{5}e^{t-4} + \frac{1}{5}e^{6(t-4)}\right]\mathscr{U}(t - 4).$$

7.6 | Systems of Linear Differential Equations

The terminology and concepts listed below provide an outline of the main ideas encountered in this section. These can be useful when preparing for a quiz or test.

Terminology and Concepts

- coupled spring systems
- doubled pendulum

The basic skills listed below summarize the more mechanical types of problems encountered in the exercise set for this section.

Basic Skills

- use Laplace transform methods to solve a linear system of differential equations
- solve algebraic systems of equations by elimination

3. Taking the Laplace transform of the system gives

$$s\mathscr{L}\{x\} + 1 = \mathscr{L}\{x\} - 2\mathscr{L}\{y\}$$

$$s\mathscr{L}\{y\} - 2 = 5\mathscr{L}\{x\} - \mathscr{L}\{y\}$$

so that

$$\mathscr{L}\{x\} = \frac{-s - 5}{s^2 + 9} = -\frac{s}{s^2 + 9} - \frac{5}{3}\frac{3}{s^2 + 9}$$

and

$$x = -\cos 3t - \frac{5}{3}\sin 3t.$$

Then

$$y = \frac{1}{2}x - \frac{1}{2}x' = 2\cos 3t - \frac{7}{3}\sin 3t.$$

6. Taking the Laplace transform of the system gives

$$(s + 1)\mathscr{L}\{x\} - (s - 1)\mathscr{L}\{y\} = -1$$

$$s\mathscr{L}\{x\} + (s + 2)\mathscr{L}\{y\} = 1$$

so that

$$\mathscr{L}\{y\} = \frac{s + 1/2}{s^2 + s + 1} = \frac{s + 1/2}{(s + 1/2)^2 + (\sqrt{3}/2)^2}$$

and

$$\mathscr{L}\{x\} = \frac{-3/2}{s^2 + s + 1} = -\sqrt{3}\frac{\sqrt{3}/2}{(s + 1/2)^2 + (\sqrt{3}/2)^2}.$$

Then

$$y = e^{-t/2}\cos\frac{\sqrt{3}}{2}t \qquad \text{and} \qquad x = -\sqrt{3}\,e^{-t/2}\sin\frac{\sqrt{3}}{2}t.$$

9. Adding the equations and then subtracting them gives

$$\frac{d^2x}{dt^2} = \frac{1}{2}t^2 + 2t$$

$$\frac{d^2y}{dt^2} = \frac{1}{2}t^2 - 2t.$$

Taking the Laplace transform of the system gives

$$\mathscr{L}\{x\} = 8\frac{1}{s} + \frac{1}{24}\frac{4!}{s^5} + \frac{1}{3}\frac{3!}{s^4}$$

and

$$\mathscr{L}\{y\} = \frac{1}{24}\frac{4!}{s^5} - \frac{1}{3}\frac{3!}{s^4}$$

so that

$$x = 8 + \frac{1}{24}t^4 + \frac{1}{3}t^3 \qquad \text{and} \qquad y = \frac{1}{24}t^4 - \frac{1}{3}t^3.$$

12. Taking the Laplace transform of the system gives

$$(s-4)\,\mathscr{L}\{x\} + 2\mathscr{L}\{y\} = \frac{2e^{-s}}{s}$$

$$-3\,\mathscr{L}\{x\} + (s+1)\,\mathscr{L}\{y\} = \frac{1}{2} + \frac{e^{-s}}{s}$$

so that

$$\mathscr{L}\{x\} = \frac{-1/2}{(s-1)(s-2)} + e^{-s}\frac{1}{(s-1)(s-2)}$$

$$= \frac{1}{2}\frac{1}{s-1} - \frac{1}{2}\frac{1}{s-2} + e^{-s}\left[-\frac{1}{s-1} + \frac{1}{s-2}\right]$$

and

$$\mathscr{L}\{y\} = \frac{e^{-s}}{s} + \frac{s/4 - 1}{(s-1)(s-2)} + e^{-s}\frac{-s/2 + 2}{(s-1)(s-2)}$$

$$= \frac{3}{4}\frac{1}{s-1} - \frac{1}{2}\frac{1}{s-2} + e^{-s}\left[\frac{1}{s} - \frac{3}{2}\frac{1}{s-1} + \frac{1}{s-2}\right].$$

Then

$$x = \frac{1}{2}e^t - \frac{1}{2}e^{2t} + \left[-e^{t-1} + e^{2(t-1)}\right]\mathscr{U}(t-1)$$

and

$$y = \frac{3}{4}e^t - \frac{1}{2}e^{2t} + \left[1 - \frac{3}{2}e^{t-1} + e^{2(t-1)}\right]\mathscr{U}(t-1).$$

15. (a) By Kirchhoff's first law we have $i_1 = i_2 + i_3$. By Kirchhoff's second law, on each loop we have
$E(t) = Ri_1 + L_1 i_2'$ and $E(t) = Ri_1 + L_2 i_3'$ or $L_1 i_2' + Ri_2 + Ri_3 = E(t)$ and $L_2 i_3' + Ri_2 + Ri_3 = E(t)$.

(b) Taking the Laplace transform of the system

$$0.01 i_2' + 5i_2 + 5i_3 = 100$$

$$0.0125 i_3' + 5i_2 + 5i_3 = 100$$

gives

$$(s + 500)\,\mathscr{L}\{i_2\} + 500\mathscr{L}\{i_3\} = \frac{10,000}{s}$$

$$400\mathscr{L}\{i_2\} + (s + 400)\,\mathscr{L}\{i_3\} = \frac{8,000}{s}$$

so that

$$\mathscr{L}\{i_3\} = \frac{8,000}{s^2 + 900s} = \frac{80}{9}\frac{1}{s} - \frac{80}{9}\frac{1}{s + 900}.$$

Then

$$i_3 = \frac{80}{9} - \frac{80}{9}e^{-900t} \qquad \text{and} \qquad i_2 = 20 - 0.0025 i_3' - i_3 = \frac{100}{9} - \frac{100}{9}e^{-900t}.$$

18. Taking the Laplace transform of the system

$$0.5 i_1' + 50 i_2 = 60$$

$$0.005 i_2' + i_2 - i_1 = 0$$

gives

$$s\,\mathscr{L}\{i_1\} + 100\,\mathscr{L}\{i_2\} = \frac{120}{s}$$

$$-200\,\mathscr{L}\{i_1\} + (s + 200)\,\mathscr{L}\{i_2\} = 0$$

so that

$$\mathscr{L}\{i_2\} = \frac{24,000}{s(s^2 + 200s + 20,000)} = \frac{6}{5}\frac{1}{s} - \frac{6}{5}\frac{s + 100}{(s + 100)^2 + 100^2} - \frac{6}{5}\frac{100}{(s + 100)^2 + 100^2}.$$

Then

$$i_2 = \frac{6}{5} - \frac{6}{5}e^{-100t}\cos 100t - \frac{6}{5}e^{-100t}\sin 100t$$

and

$$i_1 = 0.005 i_2' + i_2 = \frac{6}{5} - \frac{6}{5}e^{-100t}\cos 100t.$$

21. (a) Taking the Laplace transform of the system

$$4\theta_1'' + \theta_2'' + 8\theta_1 = 0$$

$$\theta_1'' + \theta_2'' + 2\theta_2 = 0$$

gives

$$4\left(s^2 + 2\right)\mathcal{L}\{\theta_1\} + s^2\mathcal{L}\{\theta_2\} = 3s$$

$$s^2\mathcal{L}\{\theta_1\} + \left(s^2 + 2\right)\mathcal{L}\{\theta_2\} = 0$$

so that

$$\left(3s^2 + 4\right)\left(s^2 + 4\right)\mathcal{L}\{\theta_2\} = -3s^3$$

or

$$\mathcal{L}\{\theta_2\} = \frac{1}{2}\frac{s}{s^2 + 4/3} - \frac{3}{2}\frac{s}{s^2 + 4}.$$

Then

$$\theta_2 = \frac{1}{2}\cos\frac{2}{\sqrt{3}}t - \frac{3}{2}\cos 2t \qquad \text{and} \qquad \theta_1'' = -\theta_2'' - 2\theta_2$$

so that

$$\theta_1 = \frac{1}{4}\cos\frac{2}{\sqrt{3}}t + \frac{3}{4}\cos 2t.$$

(b)

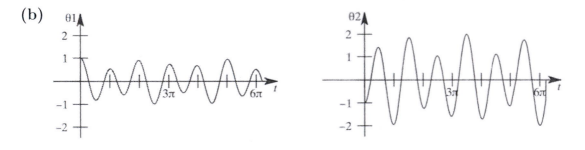

Mass m_2 has extreme displacements of greater magnitude. Mass m_1 first passes through its equilibrium position at about $t = 0.87$, and mass m_2 first passes through its equilibrium position at about $t = 0.66$. The motion of the pendulums is not periodic since $\cos(2t/\sqrt{3})$ has period $\sqrt{3}\,\pi$, $\cos 2t$ has period π, and the ratio of these periods is $\sqrt{3}$, which is not a rational number.

(c) The Lissajous curve is plotted for $0 \le t \le 30$.

(d)

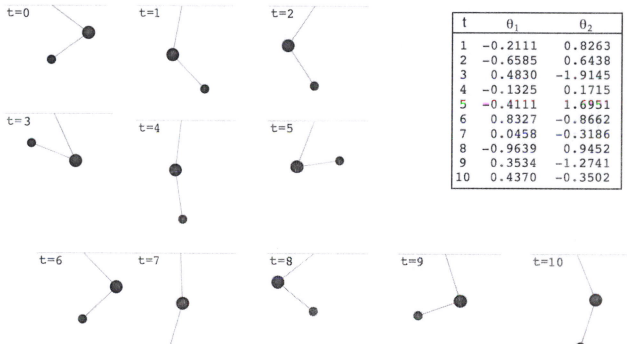

t	θ_1	θ_2
1	−0.2111	0.8263
2	−0.6585	0.6438
3	0.4830	−1.9145
4	−0.1325	0.1715
5	−0.4111	1.6951
6	0.8327	−0.8662
7	0.0458	−0.3186
8	−0.9639	0.9452
9	0.3534	−1.2741
10	0.4370	−0.3502

(e) Using a CAS to solve $\theta_1(t) = \theta_2(t)$ we see that $\theta_1 = \theta_2$ (so that the double pendulum is straight out) when t is about 0.75 seconds.

t=0.75

(f) To make a movie of the pendulum it is necessary to locate the mass in the plane as a function of time. Suppose that the upper arm is attached to the origin and that the equilibrium position lies along the negative y-axis. Then mass m_1 is at $(x,(t), y_1(t))$ and mass m_2 is at $(x_2(t), y_2(t))$, where

$$x_1(t) = 16 \sin \theta_1(t) \qquad \text{and} \qquad y_1(t) = -16 \cos \theta_1(t)$$

and

$$x_2(t) = x_1(t) + 16 \sin \theta_2(t) \qquad \text{and} \qquad y_2(t) = y_1(t) - 16 \cos \theta_2(t).$$

A reasonable movie can be constructed by letting t range from 0 to 10 in increments of 0.1 seconds.

7.R Chapter 7 in Review

3. False; consider $f(t) = t^{-1/2}$.

6. False; consider $f(t) = 1$ and $g(t) = 1$.

9. $\mathscr{L}\{\sin 2t\} = \dfrac{2}{s^2 + 4}$

12. $\mathscr{L}\{\sin 2t\, \mathscr{U}(t - \pi)\} = \mathscr{L}\{\sin 2(t - \pi)\, \mathscr{U}(t - \pi)\} = \dfrac{2}{s^2 + 4}e^{-\pi s}$

15. $\mathscr{L}^{-1}\left\{\dfrac{1}{(s-5)^3}\right\} = \dfrac{1}{2}\dfrac{2}{(s-5)^3} = \dfrac{1}{2}t^2 e^{5t}$

18. $\mathscr{L}^{-1}\left\{\dfrac{1}{s^2}e^{-5s}\right\} = (t-5)\,\mathscr{U}(t-5)$

21. $\mathscr{L}\{e^{-5t}\}$ exists for $s > -5$.

24. $\mathscr{L}\left\{\displaystyle\int_0^t e^{a\tau} f(\tau)\, d\tau\right\} = \dfrac{1}{s}\mathscr{L}\{e^{at} f(t)\} = \dfrac{F(s-a)}{s}$, whereas

$$\mathscr{L}\left\{e^{at}\int_0^t f(\tau)\, d\tau\right\} = \mathscr{L}\left\{\int_0^t f(\tau)\, d\tau\right\}\Bigg|_{s \to s-a} = \dfrac{F(s)}{s}\Bigg|_{s \to s-a} = \dfrac{F(s-a)}{s-a}.$$

27. $f(t - t_0)\,\mathscr{U}(t - t_0)$

30. $f(t) = \sin t\,\mathscr{U}(t - \pi) - \sin t\,\mathscr{U}(t - 3\pi) = -\sin(t - \pi)\,\mathscr{U}(t - \pi) + \sin(t - 3\pi)\,\mathscr{U}(t - 3\pi)$

$$\mathscr{L}\{f(t)\} = -\frac{1}{s^2 + 1}e^{-\pi s} + \frac{1}{s^2 + 1}e^{-3\pi s}$$

$$\mathscr{L}\{e^t f(t)\} = -\frac{1}{(s-1)^2 + 1}e^{-\pi(s-1)} + \frac{1}{(s-1)^2 + 1}e^{-3\pi(s-1)}$$

33. Taking the Laplace transform of the differential equation we obtain

$$\mathscr{L}\{y\} = \frac{5}{(s-1)^2} + \frac{1}{2}\frac{2}{(s-1)^3}$$

so that

$$y = 5te^t + \frac{1}{2}t^2 e^t.$$

36. Taking the Laplace transform of the differential equation we obtain

$$\mathscr{L}\{y\} = \frac{s^3 + 2}{s^3(s - 5)} - \frac{2 + 2s + s^2}{s^3(s - 5)}e^{-s}$$

$$= -\frac{2}{125}\frac{1}{s} - \frac{2}{25}\frac{1}{s^2} - \frac{1}{5}\frac{2}{s^3} + \frac{127}{125}\frac{1}{s - 5} - \left[-\frac{37}{125}\frac{1}{s} - \frac{12}{25}\frac{1}{s^2} - \frac{1}{5}\frac{2}{s^3} + \frac{37}{125}\frac{1}{s - 5}\right]e^{-s}$$

so that

$$y = -\frac{2}{125} - \frac{2}{25}t - \frac{1}{5}t^2 + \frac{127}{125}e^{5t} - \left[-\frac{37}{125} - \frac{12}{25}(t - 1) - \frac{1}{5}(t - 1)^2 + \frac{37}{125}e^{5(t-1)}\right]\mathscr{U}(t - 1).$$

39. Taking the Laplace transform of the integral equation we obtain

$$\mathscr{L}\{y\} = \frac{1}{s} + \frac{1}{s^2} + \frac{1}{2}\frac{2}{s^3}$$

so that

$$y(t) = 1 + t + \frac{1}{2}t^2.$$

42. Taking the Laplace transform of the system gives

$$s^2 \mathscr{L}\{x\} + s^2 \mathscr{L}\{y\} = \frac{1}{s-2}$$

$$2s\mathscr{L}\{x\} + s^2 \mathscr{L}\{y\} = -\frac{1}{s-2}$$

so that

$$\mathscr{L}\{x\} = \frac{2}{s(s-2)^2} = \frac{1}{2}\frac{1}{s} - \frac{1}{2}\frac{1}{s-2} + \frac{1}{(s-2)^2}$$

and

$$\mathscr{L}\{y\} = \frac{-s-2}{s^2(s-2)^2} = -\frac{3}{4}\frac{1}{s} - \frac{1}{2}\frac{1}{s^2} + \frac{3}{4}\frac{1}{s-2} - \frac{1}{(s-2)^2}.$$

Then

$$x = \frac{1}{2} - \frac{1}{2}e^{2t} + te^{2t} \quad \text{and} \quad y = -\frac{3}{4} - \frac{1}{2}t + \frac{3}{4}e^{2t} - te^{2t}.$$

45. Taking the Laplace transform of the given differential equation we obtain

$$\mathscr{L}\{y\} = \frac{2w_0}{EIL}\left(\frac{L}{48}\cdot\frac{4!}{s^5} - \frac{1}{120}\cdot\frac{5!}{s^6} + \frac{1}{120}\cdot\frac{5!}{s^6}e^{-sL/2}\right) + \frac{c_1}{2}\cdot\frac{2!}{s^3} + \frac{c_2}{6}\cdot\frac{3!}{s^4}$$

so that

$$y = \frac{2w_0}{EIL}\left[\frac{L}{48}x^4 - \frac{1}{120}x^5 + \frac{1}{120}\left(x - \frac{L}{2}\right)^5 \mathscr{U}\left(x - \frac{L}{2}\right) + \frac{c_1}{2}x^2 + \frac{c_2}{6}x^3\right]$$

where $y''(0) = c_1$ and $y'''(0) = c_2$. Using $y''(L) = 0$ and $y'''(L) = 0$ we find

$$c_1 = w_0 L^2/24EI, \qquad c_2 = -w_0 L/4EI.$$

Hence

$$y = \frac{w_0}{12EIL}\left[-\frac{1}{5}x^5 + \frac{L}{2}x^4 - \frac{L^2}{2}x^3 + \frac{L^3}{4}x^2 + \frac{1}{5}\left(x - \frac{L}{2}\right)^5 \mathscr{U}\left(x - \frac{L}{2}\right)\right].$$

48. (a) We will find the first two times for which $x'(t) = 0$ and then obtain the rest of the times using periodicity of $x'(t)$. The solution of

$$x'' + \omega^2 x = F, \quad x(0) = x_0, \quad x'(0) = 0$$

is

$$x(t) = \left(x_0 - \frac{F}{\omega^2}\right)\cos\omega t + \frac{F}{\omega^2} \quad \text{so} \quad x'(t) = \left(x_0 - \frac{F}{\omega^2}\right)(-\omega\sin\omega t).$$

The latter equation is 0 when $t = \pi/\omega$. The next initial-value problem is then

$$x'' + \omega^2 x = -F, \quad x\left(\frac{\pi}{\omega}\right) = \frac{2F}{\omega^2} - x_0, \quad x'\left(\frac{\pi}{\omega}\right) = 0,$$

where $\pi/\omega = T/2$. From the solution of this problem,

$$x(t) = \left(x_0 - \frac{3F}{\omega^2}\right)\cos\omega t - \frac{F}{\omega^2},$$

we see from

$$x'(t) = \left(x_0 - \frac{3F}{\omega^2}\right)(-\omega\sin\omega t) = 0$$

that $t = 2\pi/\omega = T$. Since $x'(t)$ has period $T = 2\pi/\omega$ we can see that $x'(t) = 0$ at the times

$$\frac{\pi}{\omega} + \frac{2\pi}{\omega}k \quad \text{and} \quad \frac{2\pi}{\omega} + \frac{2\pi}{\omega}k, \quad \text{for} \ \ k = 0, 1, 2, \ldots$$

which are

$$0, \ \frac{\pi}{\omega}, \ \frac{2\pi}{\omega}, \frac{3\pi}{\omega}, \ \ldots \quad \text{or} \quad 0, \ \frac{t}{2}, \ T, \ \frac{3T}{2}, \ \ldots.$$

(b) There is no motion unless the initial displacement is such that the force of the spring is greater than the force due to friction. That is,

$$k|x_0| > f_k \quad \text{or} \quad \frac{k}{m}|x_0| > \frac{f_k}{m} \quad \text{or} \quad \omega^2|x_0| > F.$$

(c) From part **(b)**, an initial displacement $x(0) = x_0$ for which $|x_0| > F/\omega^2$ will result in motion. On the other hand, an initial displacement x_0 for which $|x_0| \leq F/\omega^2$ will be insufficient to overcome the force of friction and the system will be "dead".

(d) The system can be described by the initial-value problem

$$x'' + \omega^2 x = g(t), \qquad x(0) = x_0, \ x'(0) = 0,$$

where g is a version of the meander function shown in Figure 7.4.6 in the text. In this case the amplitude of the function is F instead of 1, and the length of each line segment is $T/2$ rather than a. Then

$$\mathcal{L}\{x''\} + \omega^2\mathcal{L}\{x\} = \mathcal{L}\{g\},$$
$$s^2 X(s) - sx_0 + \omega^2 X(s) = G(s),$$

and

$$X(s) = \frac{sx_0}{s^2 + \omega^2} + \frac{1}{s^2 + \omega^2} G(s).$$

Now, using Problem 49 in Section 7.4 with F instead of 1 and $a = T/2$, we have

$$\mathcal{L}\{g(t)\} = G(s) = \frac{F}{s} \frac{1 - e^{-sT/2}}{1 + e^{-sT/2}}$$

$$= \frac{F}{s}[1 - 2e^{-sT/2} + 2e^{-sT} - 2e^{-3sT/2} + \cdots].$$

Then

$$X(s) = x_0 \frac{s}{s^2 + \omega^2} + \frac{F}{s} \frac{1}{s^2 + \omega^2}[1 - 2e^{-sT/2} + 2e^{-sT} - 2e^{-3sT/2} + \cdots]$$

$$= x_0 \frac{s}{s^2 + \omega^2} + \frac{F}{\omega^2}\left[\frac{1}{s} - \frac{s}{s^2 + \omega^2}\right][1 - 2e^{-sT/2} + 2e^{-sT} - 2e^{-3sT/2} + \cdots]$$

and

$$x(t) = x_0 \cos\omega t + \frac{F}{\omega^2}(1 - \cos\omega t) - \frac{2F}{\omega^2}\left[\mathcal{U}\left(t - \frac{T}{2}\right) - \cos\omega\left(t - \frac{T}{2}\right)\mathcal{U}\left(t - \frac{T}{2}\right)\right]$$

$$+ \frac{2F}{\omega^2}[\mathcal{U}(t - T) - \cos\omega(t - T)\mathcal{U}(t - T)]$$

$$- \frac{2F}{\omega^2}\left[\mathcal{U}\left(t - \frac{3T}{2}\right) - \cos\omega\left(t - \frac{3T}{2}\right)\mathcal{U}\left(t - \frac{3T}{2}\right)\right] + \cdots$$

or

$$x(t) = \begin{cases} x_0 \cos\omega t + \dfrac{F}{\omega^2}(1 - \cos\omega t), & 0 \leq t < T/2 \\[3mm] x_0 \cos\omega t + \dfrac{F}{\omega^2}(1 - \cos\omega t) - \dfrac{2F}{\omega^2}\left[1 - \cos\omega\left(t - \dfrac{T}{2}\right)\right], & T/2 \leq t < T \\[3mm] x_0 \cos\omega t + \dfrac{F}{\omega^2}(1 - \cos\omega t) - \dfrac{2F}{\omega^2}\left[1 - \cos\omega\left(t - \dfrac{T}{2}\right)\right] \\[3mm] \qquad\qquad\qquad + \dfrac{2F}{\omega^2}[1 - \cos\omega(t - T)], & T \leq t < 3T/2 \\[3mm] \qquad\qquad\vdots \end{cases}$$

(e) The solutions from Problem 28 in Chapter 5 in Review are

$$x(t) = \begin{cases} 4.5 \cos t + 1, & 0 \leq t < \pi \\ 2.5 \cos t - 1, & \pi \leq t < 2\pi. \end{cases}$$

On the interval $[0, 2\pi)$ the solution is

$$x(t) = \begin{cases} x_0 \cos\omega t + \dfrac{F}{\omega^2}(1 - \cos\omega t), & 0 \leq t < T/2 \\[3mm] x_0 \cos\omega t + \dfrac{F}{\omega^2}(1 - \cos\omega t) - \dfrac{2F}{\omega^2}\left[1 - \cos\omega\left(t - \dfrac{T}{2}\right)\right], & T/2 \leq t < T. \end{cases}$$

We let $m = 1$, $k = 1$, $f_k = 1$, and $x_0 = 5.5$. Then, when $T = 2\pi$, the foregoing becomes

$$x(t) = \begin{cases} 5.5\cos t + (1 - \cos t), & 0 \le t < \pi \\ 5.5\cos t + (1 - \cos t) - 2\big[1 - \cos(t - \pi)\big], & \pi \le t < 2\pi. \end{cases}$$

Simplifying this we get

$$x(t) = \begin{cases} 4.5\cos t + 1, & 0 \le t < \pi \\ 2.5\cos t - 1, & \pi \le t < 2\pi. \end{cases}$$

(f) At $t = T/2$, $x(T/2) = -x_0 + 2F/\omega^2$. At $t = T$,

$$x(T) = x_0 - 4F/\omega^2 = (x_0 - 2F/\omega^2) - 2F/\omega^2 = -x(T/2) - 2F/\omega^2.$$

At $t = 3T/2$, $x(3T/2) = -x_0 + 6F/\omega^2 = -x(T) + 2F/\omega^2$. We see in general that each successive oscillation is $2F/\omega^2$ shorter than the preceding one.

(g) The system will stay in motion until the oscillations bring the mass within the "dead zone" at which time the motion ceases.

8 Systems of Linear First-Order Differential Equations

The terminology and concepts listed below provide an outline of the main ideas encountered in this section. These can be useful when preparing for a quiz or test.

Terminology and Concepts

- linear system of first-order DEs

- homogeneous linear system of DEs

- matrix form of a linear system of DEs

- superposition principle

- linear independence of a set of solutions

- Wronskian

- fundamental set of solutions

- general solution of a linear system of DEs

- particular solution of a nonhomogeneous linear system of DEs

- complementary function of a homogeneous linear system of DEs

The basic skills listed below summarize the more mechanical types of problems encountered in the exercise set for this section.

Basic Skills

- express a linear first-order system in both normal form and matrix form

- determine whether a set of vectors forms a fundamental set

3. Let $\mathbf{X} = \begin{pmatrix} x \\ y \\ z \end{pmatrix}$. Then $\mathbf{X}' = \begin{pmatrix} -3 & 4 & -9 \\ 6 & -1 & 0 \\ 10 & 4 & 3 \end{pmatrix} \mathbf{X}$.

6. Let $\mathbf{X} = \begin{pmatrix} x \\ y \\ z \end{pmatrix}$. Then $\mathbf{X}' = \begin{pmatrix} -3 & 4 & 0 \\ 5 & 9 & 0 \\ 0 & 1 & 6 \end{pmatrix} \mathbf{X} + \begin{pmatrix} e^{-t}\sin 2t \\ 4e^{-t}\cos 2t \\ -e^{-t} \end{pmatrix}$.

9. $\dfrac{dx}{dt} = x - y + 2z + e^{-t} - 3t; \quad \dfrac{dy}{dt} = 3x - 4y + z + 2e^{-t} + t; \quad \dfrac{dz}{dt} = -2x + 5y + 6z + 2e^{-t} - t$

12. Since

$$\mathbf{X}' = \begin{pmatrix} 5\cos t - 5\sin t \\ 2\cos t - 4\sin t \end{pmatrix} e^t \quad \text{and} \quad \begin{pmatrix} -2 & 5 \\ -2 & 4 \end{pmatrix} \mathbf{X} = \begin{pmatrix} 5\cos t - 5\sin t \\ 2\cos t - 4\sin t \end{pmatrix} e^t$$

we see that

$$\mathbf{X}' = \begin{pmatrix} -2 & 5 \\ -2 & 4 \end{pmatrix} \mathbf{X}.$$

15. Since

$$\mathbf{X}' = \begin{pmatrix} 0 \\ 0 \\ 0 \end{pmatrix} \quad \text{and} \quad \begin{pmatrix} 1 & 2 & 1 \\ 6 & -1 & 0 \\ -1 & -2 & -1 \end{pmatrix} \mathbf{X} = \begin{pmatrix} 0 \\ 0 \\ 0 \end{pmatrix}$$

we see that

$$\mathbf{X}' = \begin{pmatrix} 1 & 2 & 1 \\ 6 & -1 & 0 \\ -1 & -2 & -1 \end{pmatrix} \mathbf{X}.$$

18. Yes, since $W(\mathbf{X}_1, \mathbf{X}_2) = 8e^{2t} \neq 0$ the set $\mathbf{X}_1, \mathbf{X}_2$ is linearly independent on $-\infty < t < \infty$.

21. Since

$$\mathbf{X}_p' = \begin{pmatrix} 2 \\ -1 \end{pmatrix} \quad \text{and} \quad \begin{pmatrix} 1 & 4 \\ 3 & 2 \end{pmatrix} \mathbf{X}_p + \begin{pmatrix} 2 \\ -4 \end{pmatrix} t + \begin{pmatrix} -7 \\ -18 \end{pmatrix} = \begin{pmatrix} 2 \\ -1 \end{pmatrix}$$

we see that

$$\mathbf{X}_p' = \begin{pmatrix} 1 & 4 \\ 3 & 2 \end{pmatrix} \mathbf{X}_p + \begin{pmatrix} 2 \\ -4 \end{pmatrix} t + \begin{pmatrix} -7 \\ -18 \end{pmatrix}.$$

24. Since

$$\mathbf{X}_p' = \begin{pmatrix} 3\cos 3t \\ 0 \\ -3\sin 3t \end{pmatrix} \quad \text{and} \quad \begin{pmatrix} 1 & 2 & 3 \\ -4 & 2 & 0 \\ -6 & 1 & 0 \end{pmatrix} \mathbf{X}_p + \begin{pmatrix} -1 \\ 4 \\ 3 \end{pmatrix} \sin 3t = \begin{pmatrix} 3\cos 3t \\ 0 \\ -3\sin 3t \end{pmatrix}$$

we see that

$$\mathbf{X}_p' = \begin{pmatrix} 1 & 2 & 3 \\ -4 & 2 & 0 \\ -6 & 1 & 0 \end{pmatrix} \mathbf{X}_p + \begin{pmatrix} -1 \\ 4 \\ 3 \end{pmatrix} \sin 3t.$$

8.2 | Homogeneous Linear Systems

The terminology and concepts listed below provide an outline of the main ideas encountered in this section. These can be useful when preparing for a quiz or test.

Terminology and Concepts

- characteristic equation of a matrix
- eigenvalues of a matrix
- eigenvector of a matrix corresponding to an eigenvalue
- general solution of a homogeneous linear system of DEs with constant coefficients
- phase plane
- trajectory
- phase portrait of a linear system of DEs
- repeller
- attractor
- multiplicity of an eigenvalue

The basic skills listed below summarize the more mechanical types of problems encountered in the exercise set for this section.

Basic Skills

- find the general solution of a homogeneous linear system of DEs with constant coefficients when the eigenvalues are real and distinct, real and repeated, or complex

Use of Computers
To find eigenvalues and eigenvectors of a matrix using *Mathematica* or *Maple*, we can input the matrix by rows. For example,

$$m = \begin{pmatrix} -4 & 1 & 1 \\ 1 & 5 & -1 \\ 0 & 1 & -3 \end{pmatrix}$$

is the matrix of the coefficients of the linear system (6) in Example 2 in the text. Using *Mathematica*, we can write **m** as

m={{-4, 1, 1}, {1, 5, -1}, {0, 1, -3}}

The commands **Eigenvalues[m]** and **Eigenvectors[m]** given in sequence yield

$$[-4, -3, 5] \quad \text{and} \quad \{\{-10, -1, 1\}, \{1, 0, 1\}, \{1, 8, 1\}\}.$$

Translated this means $\lambda = -4$ is an eigenvalue with corresponding eigenvector

$$\mathbf{K} = \begin{pmatrix} -10 \\ -1 \\ 1 \end{pmatrix},$$

and so on. In *Mathematica* eigenvalues and eigenvectors can be obtained at the same time using the single command **Eigensystem[m]**. In *Maple*, the matrix **m** is written

```
m:=matrix(3,3,[-4,1,1,1,5,-1,0,1,-3])
```

The commands for eigenvalues and eigenvectors are, respectively, `eigenvals(m)` and `eigenvects(m)`.

3. The system is

$$\mathbf{X}' = \begin{pmatrix} -4 & 2 \\ -5/2 & 2 \end{pmatrix} \mathbf{X}$$

and $\det(\mathbf{A} - \lambda\mathbf{I}) = (\lambda - 1)(\lambda + 3) = 0$. For $\lambda_1 = 1$ we obtain

$$\begin{pmatrix} -5 & 2 & | & 0 \\ -5/2 & 1 & | & 0 \end{pmatrix} \implies \begin{pmatrix} -5 & 2 & | & 0 \\ 0 & 0 & | & 0 \end{pmatrix} \quad \text{so that} \quad \mathbf{K}_1 = \begin{pmatrix} 2 \\ 5 \end{pmatrix}.$$

For $\lambda_2 = -3$ we obtain

$$\begin{pmatrix} -1 & 2 & | & 0 \\ -5/2 & 5 & | & 0 \end{pmatrix} \implies \begin{pmatrix} -1 & 2 & | & 0 \\ 0 & 0 & | & 0 \end{pmatrix} \quad \text{so that} \quad \mathbf{K}_2 = \begin{pmatrix} 2 \\ 1 \end{pmatrix}.$$

Then

$$\mathbf{X} = c_1 \begin{pmatrix} 2 \\ 5 \end{pmatrix} e^t + c_2 \begin{pmatrix} 2 \\ 1 \end{pmatrix} e^{-3t}.$$

6. The system is

$$\mathbf{X}' = \begin{pmatrix} -6 & 2 \\ -3 & 1 \end{pmatrix} \mathbf{X}$$

and $\det(\mathbf{A} - \lambda\mathbf{I}) = \lambda(\lambda + 5) = 0$. For $\lambda_1 = 0$ we obtain

$$\begin{pmatrix} -6 & 2 & | & 0 \\ -3 & 1 & | & 0 \end{pmatrix} \implies \begin{pmatrix} 1 & -1/3 & | & 0 \\ 0 & 0 & | & 0 \end{pmatrix} \quad \text{so that} \quad \mathbf{K}_1 = \begin{pmatrix} 1 \\ 3 \end{pmatrix}.$$

For $\lambda_2 = -5$ we obtain

$$\begin{pmatrix} -1 & 2 & | & 0 \\ -3 & 6 & | & 0 \end{pmatrix} \implies \begin{pmatrix} 1 & -2 & | & 0 \\ 0 & 0 & | & 0 \end{pmatrix} \quad \text{so that} \quad \mathbf{K}_2 = \begin{pmatrix} 2 \\ 1 \end{pmatrix}.$$

Then

$$\mathbf{X} = c_1 \begin{pmatrix} 1 \\ 3 \end{pmatrix} + c_2 \begin{pmatrix} 2 \\ 1 \end{pmatrix} e^{-5t}.$$

9. We have $\det(\mathbf{A} - \lambda\mathbf{I}) = -(\lambda + 1)(\lambda - 3)(\lambda + 2) = 0$. For $\lambda_1 = -1$, $\lambda_2 = 3$, and $\lambda_3 = -2$ we obtain

$$\mathbf{K}_1 = \begin{pmatrix} -1 \\ 0 \\ 1 \end{pmatrix}, \quad \mathbf{K}_2 = \begin{pmatrix} 1 \\ 4 \\ 3 \end{pmatrix}, \quad \text{and} \quad \mathbf{K}_3 = \begin{pmatrix} 1 \\ -1 \\ 3 \end{pmatrix},$$

so that

$$\mathbf{X} = c_1 \begin{pmatrix} -1 \\ 0 \\ 1 \end{pmatrix} e^{-t} + c_2 \begin{pmatrix} 1 \\ 4 \\ 3 \end{pmatrix} e^{3t} + c_3 \begin{pmatrix} 1 \\ -1 \\ 3 \end{pmatrix} e^{-2t}.$$

12. We have $\det(\mathbf{A} - \lambda\mathbf{I}) = (\lambda - 3)(\lambda + 5)(6 - \lambda) = 0$. For $\lambda_1 = 3$, $\lambda_2 = -5$, and $\lambda_3 = 6$ we obtain

$$\mathbf{K}_1 = \begin{pmatrix} 1 \\ 1 \\ 0 \end{pmatrix}, \quad \mathbf{K}_2 = \begin{pmatrix} 1 \\ -1 \\ 0 \end{pmatrix}, \quad \text{and} \quad \mathbf{K}_3 = \begin{pmatrix} 2 \\ -2 \\ 11 \end{pmatrix},$$

so that

$$\mathbf{X} = c_1 \begin{pmatrix} 1 \\ 1 \\ 0 \end{pmatrix} e^{3t} + c_2 \begin{pmatrix} 1 \\ -1 \\ 0 \end{pmatrix} e^{-5t} + c_3 \begin{pmatrix} 2 \\ -2 \\ 11 \end{pmatrix} e^{6t}.$$

15. $\mathbf{X} = c_1 \begin{pmatrix} 0.382175 \\ 0.851161 \\ 0.359815 \end{pmatrix} e^{8.58979t} + c_2 \begin{pmatrix} 0.405188 \\ -0.676043 \\ 0.615458 \end{pmatrix} e^{2.25684t} + c_3 \begin{pmatrix} -0.923562 \\ -0.132174 \\ 0.35995 \end{pmatrix} e^{-0.0466321t}$

18. In Problem 2, letting $c_1 = 1$ and $c_2 = 0$ we get $x = -2e^t$, $y = e^t$. Eliminating the parameter we find $y = -\frac{1}{2}x$, $x < 0$. When $c_1 = -1$ and $c_2 = 0$ we find $y = -\frac{1}{2}x$, $x > 0$. Letting $c_1 = 0$ and $c_2 = 1$ we get $x = e^{4t}$, $y = e^{4t}$. Eliminating the parameter we find $y = x$, $x > 0$. When $c_1 = 0$ and $c_2 = -1$ we find $y = x$, $x < 0$.

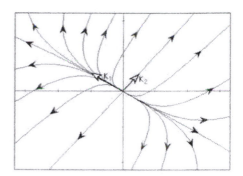

In Problem 4, letting $c_1 = 1$ and $c_2 = 0$ we get $x = -2e^{-7t/2}$, $y = e^{-7t/2}$. Eliminating the parameter we find $y = -\frac{1}{2}x$, $x < 0$. When $c_1 = -1$ and $c_2 = 0$ we find $y = -\frac{1}{2}x$, $x > 0$. Letting $c_1 = 0$ and $c_2 = 1$ we get $x = 4e^{-t}$, $y = 3e^{-t}$. Eliminating the parameter we find $y = \frac{3}{4}x$, $x > 0$. When $c_1 = 0$ and $c_2 = -1$ we find $y = \frac{3}{4}x$, $x < 0$.

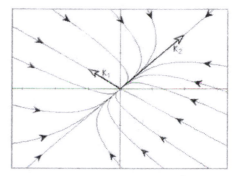

21. We have $\det(\mathbf{A} - \lambda\mathbf{I}) = (\lambda - 2)^2 = 0$. For $\lambda_1 = 2$ we obtain

$$\mathbf{K} = \begin{pmatrix} 1 \\ 1 \end{pmatrix}.$$

A solution of $(\mathbf{A} - \lambda_1\mathbf{I})\mathbf{P} = \mathbf{K}$ is

$$\mathbf{P} = \begin{pmatrix} -1/3 \\ 0 \end{pmatrix}$$

so that

$$\mathbf{X} = c_1 \begin{pmatrix} 1 \\ 1 \end{pmatrix} e^{2t} + c_2 \left[\begin{pmatrix} 1 \\ 1 \end{pmatrix} te^{2t} + \begin{pmatrix} -1/3 \\ 0 \end{pmatrix} e^{2t} \right].$$

24. We have $\det(\mathbf{A} - \lambda\mathbf{I}) = (\lambda - 8)(\lambda + 1)^2 = 0$. For $\lambda_1 = 8$ we obtain

$$\mathbf{K}_1 = \begin{pmatrix} 2 \\ 1 \\ 2 \end{pmatrix}.$$

For $\lambda_2 = -1$ we obtain

$$\mathbf{K}_2 = \begin{pmatrix} 0 \\ -2 \\ 1 \end{pmatrix} \quad \text{and} \quad \mathbf{K}_3 = \begin{pmatrix} 1 \\ -2 \\ 0 \end{pmatrix}.$$

Then

$$\mathbf{X} = c_1 \begin{pmatrix} 2 \\ 1 \\ 2 \end{pmatrix} e^{8t} + c_2 \begin{pmatrix} 0 \\ -2 \\ 1 \end{pmatrix} e^{-t} + c_3 \begin{pmatrix} 1 \\ -2 \\ 0 \end{pmatrix} e^{-t}.$$

27. We have $\det(\mathbf{A} - \lambda\mathbf{I}) = -(\lambda - 1)^3 = 0$. For $\lambda_1 = 1$ we obtain

$$\mathbf{K} = \begin{pmatrix} 0 \\ 1 \\ 1 \end{pmatrix}.$$

Solutions of $(\mathbf{A} - \lambda_1\mathbf{I})\mathbf{P} = \mathbf{K}$ and $(\mathbf{A} - \lambda_1\mathbf{I})\mathbf{Q} = \mathbf{P}$ are

$$\mathbf{P} = \begin{pmatrix} 0 \\ 1 \\ 0 \end{pmatrix} \quad \text{and} \quad \mathbf{Q} = \begin{pmatrix} 1/2 \\ 0 \\ 0 \end{pmatrix}$$

so that

$$\mathbf{X} = c_1 \begin{pmatrix} 0 \\ 1 \\ 1 \end{pmatrix} e^t + c_2 \left[\begin{pmatrix} 0 \\ 1 \\ 1 \end{pmatrix} te^t + \begin{pmatrix} 0 \\ 1 \\ 0 \end{pmatrix} e^t \right] + c_3 \left[\begin{pmatrix} 0 \\ 1 \\ 1 \end{pmatrix} \frac{t^2}{2}e^t + \begin{pmatrix} 0 \\ 1 \\ 0 \end{pmatrix} te^t + \begin{pmatrix} 1/2 \\ 0 \\ 0 \end{pmatrix} e^t \right].$$

30. We have $\det(\mathbf{A} - \lambda\mathbf{I}) = -(\lambda + 1)(\lambda - 1)^2 = 0$. For $\lambda_1 = -1$ we obtain

$$\mathbf{K}_1 = \begin{pmatrix} -1 \\ 0 \\ 1 \end{pmatrix}.$$

For $\lambda_2 = 1$ we obtain

$$\mathbf{K}_2 = \begin{pmatrix} 1 \\ 0 \\ 1 \end{pmatrix} \quad \text{and} \quad \mathbf{K}_3 = \begin{pmatrix} 0 \\ 1 \\ 0 \end{pmatrix}$$

so that

$$\mathbf{X} = c_1 \begin{pmatrix} -1 \\ 0 \\ 1 \end{pmatrix} e^{-t} + c_2 \begin{pmatrix} 1 \\ 0 \\ 1 \end{pmatrix} e^t + c_3 \begin{pmatrix} 0 \\ 1 \\ 0 \end{pmatrix} e^t.$$

If

$$\mathbf{X}(0) = \begin{pmatrix} 1 \\ 2 \\ 5 \end{pmatrix}$$

then $c_1 = 2$, $c_2 = 3$, and $c_3 = 2$.

33. We have $\det(\mathbf{A} - \lambda\mathbf{I}) = \lambda^2 - 8\lambda + 17 = 0$. For $\lambda_1 = 4 + i$ we obtain

$$\mathbf{K}_1 = \begin{pmatrix} 2 + i \\ 5 \end{pmatrix}$$

so that

$$\mathbf{X}_1 = \begin{pmatrix} 2 + i \\ 5 \end{pmatrix} e^{(4+i)t} = \begin{pmatrix} 2\cos t - \sin t \\ 5\cos t \end{pmatrix} e^{4t} + i \begin{pmatrix} \cos t + 2\sin t \\ 5\sin t \end{pmatrix} e^{4t}$$

Then

$$\mathbf{X} = c_1 \begin{pmatrix} 2\cos t - \sin t \\ 5\cos t \end{pmatrix} e^{4t} + c_2 \begin{pmatrix} \cos t + 2\sin t \\ 5\sin t \end{pmatrix} e^{4t}.$$

36. We have $\det(\mathbf{A} - \lambda\mathbf{I}) = \lambda^2 - 10\lambda + 34 = 0$. For $\lambda_1 = 5 + 3i$ we obtain

$$\mathbf{K}_1 = \begin{pmatrix} 1 - 3i \\ 2 \end{pmatrix}$$

so that

$$\mathbf{X}_1 = \begin{pmatrix} 1 - 3i \\ 2 \end{pmatrix} e^{(5+3i)t} = \begin{pmatrix} \cos 3t + 3\sin 3t \\ 2\cos 3t \end{pmatrix} e^{5t} + i \begin{pmatrix} \sin 3t - 3\cos 3t \\ 2\sin 3t \end{pmatrix} e^{5t}.$$

Then

$$\mathbf{X} = c_1 \begin{pmatrix} \cos 3t + 3\sin 3t \\ 2\cos 3t \end{pmatrix} e^{5t} + c_2 \begin{pmatrix} \sin 3t - 3\cos 3t \\ 2\sin 3t \end{pmatrix} e^{5t}.$$

39. We have $\det(\mathbf{A} - \lambda\mathbf{I}) = -\lambda\left(\lambda^2 + 1\right) = 0$. For $\lambda_1 = 0$ we obtain

$$\mathbf{K}_1 = \begin{pmatrix} 1 \\ 0 \\ 0 \end{pmatrix}.$$

For $\lambda_2 = i$ we obtain

$$\mathbf{K}_2 = \begin{pmatrix} -i \\ i \\ 1 \end{pmatrix}$$

so that

$$\mathbf{X}_2 = \begin{pmatrix} -i \\ i \\ 1 \end{pmatrix} e^{it} = \begin{pmatrix} \sin t \\ -\sin t \\ \cos t \end{pmatrix} + i \begin{pmatrix} -\cos t \\ \cos t \\ \sin t \end{pmatrix}.$$

Then

$$\mathbf{X} = c_1 \begin{pmatrix} 1 \\ 0 \\ 0 \end{pmatrix} + c_2 \begin{pmatrix} \sin t \\ -\sin t \\ \cos t \end{pmatrix} + c_3 \begin{pmatrix} -\cos t \\ \cos t \\ \sin t \end{pmatrix}.$$

42. We have $\det(\mathbf{A} - \lambda\mathbf{I}) = -(\lambda - 6)(\lambda^2 - 8\lambda + 20) = 0$. For $\lambda_1 = 6$ we obtain

$$\mathbf{K}_1 = \begin{pmatrix} 0 \\ 1 \\ 0 \end{pmatrix}.$$

For $\lambda_2 = 4 + 2i$ we obtain

$$\mathbf{K}_2 = \begin{pmatrix} -i \\ 0 \\ 2 \end{pmatrix}$$

so that

$$\mathbf{X}_2 = \begin{pmatrix} -i \\ 0 \\ 2 \end{pmatrix} e^{(4+2i)t} = \begin{pmatrix} \sin 2t \\ 0 \\ 2\cos 2t \end{pmatrix} e^{4t} + i \begin{pmatrix} -\cos 2t \\ 0 \\ 2\sin 2t \end{pmatrix} e^{4t}.$$

Then

$$\mathbf{X} = c_1 \begin{pmatrix} 0 \\ 1 \\ 0 \end{pmatrix} e^{6t} + c_2 \begin{pmatrix} \sin 2t \\ 0 \\ 2\cos 2t \end{pmatrix} e^{4t} + c_3 \begin{pmatrix} -\cos 2t \\ 0 \\ 2\sin 2t \end{pmatrix} e^{4t}.$$

45. We have $\det(\mathbf{A} - \lambda\mathbf{I}) = (1 - \lambda)(\lambda^2 + 25) = 0$. For $\lambda_1 = 1$ we obtain

$$\mathbf{K}_1 = \begin{pmatrix} 25 \\ -7 \\ 6 \end{pmatrix}.$$

For $\lambda_2 = 5i$ we obtain

$$\mathbf{K}_2 = \begin{pmatrix} 1 + 5i \\ 1 \\ 1 \end{pmatrix}$$

so that

$$\mathbf{X}_2 = \begin{pmatrix} 1 + 5i \\ 1 \\ 1 \end{pmatrix} e^{5it} = \begin{pmatrix} \cos 5t - 5\sin 5t \\ \cos 5t \\ \cos 5t \end{pmatrix} + i \begin{pmatrix} \sin 5t + 5\cos 5t \\ \sin 5t \\ \sin 5t \end{pmatrix}.$$

Then

$$\mathbf{X} = c_1 \begin{pmatrix} 25 \\ -7 \\ 6 \end{pmatrix} e^{t} + c_2 \begin{pmatrix} \cos 5t - 5\sin 5t \\ \cos 5t \\ \cos 5t \end{pmatrix} + c_3 \begin{pmatrix} \sin 5t + 5\cos 5t \\ \sin 5t \\ \sin 5t \end{pmatrix}.$$

If

$$\mathbf{X}(0) = \begin{pmatrix} 4 \\ 6 \\ -7 \end{pmatrix}$$

then $c_1 = c_2 = -1$ and $c_3 = 6$.

48. (a) Letting $x_1 = y_1$, $x_1' = y_2$, $x_2 = y_3$, and $x_2' = y_4$ we have

$$y_2' = x_1'' = -10x_1 + 4x_2 = -10y_1 + 4y_3$$

$$y_4' = x_2'' = 4x_1 - 4x_2 = 4y_1 - 4y_3.$$

The corresponding linear system is

$$y_1' = y_2$$

$$y_2' = -10y_1 + 4y_3$$

$$y_3' = y_4$$

$$y_4' = 4y_1 - 4y_3$$

or

$$\mathbf{Y}' = \begin{pmatrix} 0 & 1 & 0 & 0 \\ -10 & 0 & 4 & 0 \\ 0 & 0 & 0 & 1 \\ 4 & 0 & -4 & 0 \end{pmatrix} \mathbf{Y}.$$

Using a CAS, we find eigenvalues $\pm\sqrt{2}i$ and $\pm2\sqrt{3}i$ with corresponding eigenvectors

$$\begin{pmatrix} \mp\sqrt{2}i/4 \\ 1/2 \\ \mp\sqrt{2}i/2 \\ 1 \end{pmatrix} = \begin{pmatrix} 0 \\ 1/2 \\ 0 \\ 1 \end{pmatrix} + i \begin{pmatrix} \mp\sqrt{2}/4 \\ 0 \\ \mp\sqrt{2}/2 \\ 0 \end{pmatrix}$$

and

$$\begin{pmatrix} \pm\sqrt{3}i/3 \\ -2 \\ \mp\sqrt{3}i/6 \\ 1 \end{pmatrix} = \begin{pmatrix} 0 \\ -2 \\ 0 \\ 1 \end{pmatrix} + i \begin{pmatrix} \pm\sqrt{3}/3 \\ 0 \\ \mp\sqrt{3}/6 \\ 0 \end{pmatrix}.$$

Thus

$$
\begin{aligned}
\mathbf{Y}(t) = c_1 & \left[\begin{pmatrix} 0 \\ 1/2 \\ 0 \\ 1 \end{pmatrix} \cos \sqrt{2}\,t - \begin{pmatrix} -\sqrt{2}/4 \\ 0 \\ -\sqrt{2}/2 \\ 0 \end{pmatrix} \sin \sqrt{2}\,t \right] \\
+ c_2 & \left[\begin{pmatrix} -\sqrt{2}/4 \\ 0 \\ -\sqrt{2}/2 \\ 0 \end{pmatrix} \cos \sqrt{2}\,t + \begin{pmatrix} 0 \\ 1/2 \\ 0 \\ 1 \end{pmatrix} \sin \sqrt{2}\,t \right] \\
+ c_3 & \left[\begin{pmatrix} 0 \\ -2 \\ 0 \\ 1 \end{pmatrix} \cos 2\sqrt{3}\,t - \begin{pmatrix} \sqrt{3}/3 \\ 0 \\ -\sqrt{3}/6 \\ 0 \end{pmatrix} \sin 2\sqrt{3}\,t \right] \\
+ c_4 & \left[\begin{pmatrix} \sqrt{3}/3 \\ 0 \\ -\sqrt{3}/6 \\ 0 \end{pmatrix} \cos 2\sqrt{3}\,t + \begin{pmatrix} 0 \\ -2 \\ 0 \\ 1 \end{pmatrix} \sin 2\sqrt{3}\,t \right].
\end{aligned}
$$

The initial conditions $y_1(0) = 0$, $y_2(0) = 1$, $y_3(0) = 0$, and $y_4(0) = -1$ imply $c_1 = -\frac{2}{5}$, $c_2 = 0$, $c_3 = -\frac{3}{5}$, and $c_4 = 0$. Thus,

$$
x_1(t) = y_1(t) = -\frac{\sqrt{2}}{10} \sin \sqrt{2}\,t + \frac{\sqrt{3}}{5} \sin 2\sqrt{3}\,t
$$

$$
x_2(t) = y_3(t) = -\frac{\sqrt{2}}{5} \sin \sqrt{2}\,t - \frac{\sqrt{3}}{10} \sin 2\sqrt{3}\,t.
$$

(b) The second-order system is

$$
x_1'' = -10x_1 + 4x_2
$$

$$
x_2'' = 4x_1 - 4x_2
$$

or

$$
\mathbf{X}'' = \begin{pmatrix} -10 & 4 \\ 4 & -4 \end{pmatrix} \mathbf{X}.
$$

We assume solutions of the form $\mathbf{X} = \mathbf{V} \cos \omega t$ and $\mathbf{X} = \mathbf{V} \sin \omega t$. Since the eigenvalues are -2 and -12, $\omega_1 = \sqrt{-(-2)} = \sqrt{2}$ and $\omega_2 = \sqrt{-(-12)} = 2\sqrt{3}$. The corresponding eigenvectors are

$$
\mathbf{V}_1 = \begin{pmatrix} 1 \\ 2 \end{pmatrix} \qquad \text{and} \qquad \mathbf{V}_2 = \begin{pmatrix} -2 \\ 1 \end{pmatrix}.
$$

Then, the general solution of the system is

$$\mathbf{X} = c_1 \begin{pmatrix} 1 \\ 2 \end{pmatrix} \cos \sqrt{2}t + c_2 \begin{pmatrix} 1 \\ 2 \end{pmatrix} \sin \sqrt{2}t + c_3 \begin{pmatrix} -2 \\ 1 \end{pmatrix} \cos 2\sqrt{3}t + c_4 \begin{pmatrix} -2 \\ 1 \end{pmatrix} \sin 2\sqrt{3}t.$$

The initial conditions

$$\mathbf{X}(0) = \begin{pmatrix} 0 \\ 0 \end{pmatrix} \quad \text{and} \quad \mathbf{X}'(0) = \begin{pmatrix} 1 \\ -1 \end{pmatrix}$$

imply $c_1 = 0$, $c_2 = -\sqrt{2}/10$, $c_3 = 0$, and $c_4 = -\sqrt{3}/10$. Thus

$$x_1(t) = -\frac{\sqrt{2}}{10} \sin \sqrt{2}t + \frac{\sqrt{3}}{5} \sin 2\sqrt{3}t$$

$$x_2(t) = -\frac{\sqrt{2}}{5} \sin \sqrt{2}t - \frac{\sqrt{3}}{10} \sin 2\sqrt{3}t.$$

51. From $x = 2\cos 2t - 2\sin 2t$, $y = -\cos 2t$ we find $x + 2y = -2\sin 2t$. Then

$$(x + 2y)^2 = 4\sin^2 2t = 4(1 - \cos^2 2t) = 4 - 4\cos^2 2t = 4 - 4y^2$$

and

$$x^2 + 4xy + 4y^2 = 4 - 4y^2 \quad \text{or} \quad x^2 + 4xy + 8y^2 = 4.$$

This is a rotated conic section and, from the discriminant $b^2 - 4ac = 16 - 32 < 0$, we see that the curve is an ellipse.

8.3 Nonhomogeneous Linear Systems

The terminology and concepts listed below provide an outline of the main ideas encountered in this section. These can be useful when preparing for a quiz or test.

Terminology and Concepts

- method of undetermined coefficients for finding a particular solution of a nonhomogeneous linear system of DEs

- fundamental matrix of a linear system of DEs

- method of variation of parameters for finding a particular solution of a nonhomogeneous linear system of DEs

The basic skills listed below summarize the more mechanical types of problems encountered in the exercise set for this section.

Basic Skills

- use undetermined coefficients to solve a nonhomogeneous linear system of DEs
- use variation of parameters to solve a nonhomogeneous linear system of DEs

3. Solving

$$\det(\mathbf{A} - \lambda\mathbf{I}) = \begin{vmatrix} 1 - \lambda & 3 \\ 3 & 1 - \lambda \end{vmatrix} = \lambda^2 - 2\lambda - 8 = (\lambda - 4)(\lambda + 2) = 0$$

we obtain eigenvalues $\lambda_1 = -2$ and $\lambda_2 = 4$. Corresponding eigenvectors are

$$\mathbf{K}_1 = \begin{pmatrix} 1 \\ -1 \end{pmatrix} \quad \text{and} \quad \mathbf{K}_2 = \begin{pmatrix} 1 \\ 1 \end{pmatrix}.$$

Thus

$$\mathbf{X}_c = c_1 \begin{pmatrix} 1 \\ -1 \end{pmatrix} e^{-2t} + c_2 \begin{pmatrix} 1 \\ 1 \end{pmatrix} e^{4t}.$$

Substituting

$$\mathbf{X}_p = \begin{pmatrix} a_3 \\ b_3 \end{pmatrix} t^2 + \begin{pmatrix} a_2 \\ b_2 \end{pmatrix} t + \begin{pmatrix} a_1 \\ b_1 \end{pmatrix}$$

into the system yields

$$a_3 + 3b_3 = 2 \qquad a_2 + 3b_2 = 2a_3 \qquad a_1 + 3b_1 = a_2$$

$$3a_3 + b_3 = 0 \qquad 3a_2 + b_2 + 1 = 2b_3 \qquad 3a_1 + b_1 + 5 = b_2$$

from which we obtain $a_3 = -1/4$, $b_3 = 3/4$, $a_2 = 1/4$, $b_2 = -1/4$, $a_1 = -2$, and $b_1 = 3/4$. Then

$$\mathbf{X}(t) = c_1 \begin{pmatrix} 1 \\ -1 \end{pmatrix} e^{-2t} + c_2 \begin{pmatrix} 1 \\ 1 \end{pmatrix} e^{4t} + \begin{pmatrix} -1/4 \\ 3/4 \end{pmatrix} t^2 + \begin{pmatrix} 1/4 \\ -1/4 \end{pmatrix} t + \begin{pmatrix} -2 \\ 3/4 \end{pmatrix}.$$

6. Solving

$$\det(\mathbf{A} - \lambda\mathbf{I}) = \begin{vmatrix} -1 - \lambda & 5 \\ -1 & 1 - \lambda \end{vmatrix} = \lambda^2 + 4 = 0$$

we obtain the eigenvalues $\lambda_1 = 2i$ and $\lambda_2 = -2i$. Corresponding eigenvectors are

$$\mathbf{K}_1 = \begin{pmatrix} 5 \\ 1 + 2i \end{pmatrix} \quad \text{and} \quad \mathbf{K}_2 = \begin{pmatrix} 5 \\ 1 - 2i \end{pmatrix}.$$

Thus

$$\mathbf{X}_c = c_1 \begin{pmatrix} 5\cos 2t \\ \cos 2t - 2\sin 2t \end{pmatrix} + c_2 \begin{pmatrix} 5\sin 2t \\ 2\cos 2t + \sin 2t \end{pmatrix}.$$

Substituting

$$\mathbf{X}_p = \begin{pmatrix} a_2 \\ b_2 \end{pmatrix} \cos t + \begin{pmatrix} a_1 \\ b_1 \end{pmatrix} \sin t$$

into the system yields

$$-a_2 + 5b_2 - a_1 = 0$$

$$-a_2 + b_2 - b_1 - 2 = 0$$

$$-a_1 + 5b_1 + a_2 + 1 = 0$$

$$-a_1 + b_1 + b_2 = 0$$

from which we obtain $a_2 = -3$, $b_2 = -2/3$, $a_1 = -1/3$, and $b_1 = 1/3$. Then

$$\mathbf{X}(t) = c_1 \begin{pmatrix} 5\cos 2t \\ \cos 2t - 2\sin 2t \end{pmatrix} + c_2 \begin{pmatrix} 5\sin 2t \\ 2\cos 2t + \sin 2t \end{pmatrix} + \begin{pmatrix} -3 \\ -2/3 \end{pmatrix} \cos t + \begin{pmatrix} -1/3 \\ 1/3 \end{pmatrix} \sin t.$$

9. Solving

$$\det(\mathbf{A} - \lambda \mathbf{I}) = \begin{vmatrix} -1-\lambda & -2 \\ 3 & 4-\lambda \end{vmatrix} = \lambda^2 - 3\lambda + 2 = (\lambda-1)(\lambda-2) = 0$$

we obtain the eigenvalues $\lambda_1 = 1$ and $\lambda_2 = 2$. Corresponding eigenvectors are

$$\mathbf{K}_1 = \begin{pmatrix} 1 \\ -1 \end{pmatrix} \quad \text{and} \quad \mathbf{K}_2 = \begin{pmatrix} -4 \\ 6 \end{pmatrix}.$$

Thus

$$\mathbf{X}_c = c_1 \begin{pmatrix} 1 \\ -1 \end{pmatrix} e^t + c_2 \begin{pmatrix} -4 \\ 6 \end{pmatrix} e^{2t}.$$

Substituting

$$\mathbf{X}_p = \begin{pmatrix} a_1 \\ b_1 \end{pmatrix}$$

into the system yields

$$-a_1 - 2b_1 = -3$$

$$3a_1 + 4b_1 = -3$$

from which we obtain $a_1 = -9$ and $b_1 = 6$. Then

$$\mathbf{X}(t) = c_1 \begin{pmatrix} 1 \\ -1 \end{pmatrix} e^t + c_2 \begin{pmatrix} -4 \\ 6 \end{pmatrix} e^{2t} + \begin{pmatrix} -9 \\ 6 \end{pmatrix}.$$

Setting

$$\mathbf{X}(0) = \begin{pmatrix} -4 \\ 5 \end{pmatrix}$$

we obtain

$$c_1 - 4c_2 - 9 = -4$$

$$-c_1 + 6c_2 + 6 = 5.$$

Then $c_1 = 13$ and $c_2 = 2$ so

$$\mathbf{X}(t) = 13 \begin{pmatrix} 1 \\ -1 \end{pmatrix} e^t + 2 \begin{pmatrix} -4 \\ 6 \end{pmatrix} e^{2t} + \begin{pmatrix} -9 \\ 6 \end{pmatrix}.$$

12. From

$$\mathbf{X}' = \begin{pmatrix} 2 & -1 \\ 3 & -2 \end{pmatrix} \mathbf{X} + \begin{pmatrix} 0 \\ 4 \end{pmatrix} t$$

we obtain

$$\mathbf{X}_c = c_1 \begin{pmatrix} 1 \\ 1 \end{pmatrix} e^t + c_2 \begin{pmatrix} 1 \\ 3 \end{pmatrix} e^{-t}.$$

Then

$$\boldsymbol{\Phi} = \begin{pmatrix} e^t & e^{-t} \\ e^t & 3e^{-t} \end{pmatrix} \quad \text{and} \quad \boldsymbol{\Phi}^{-1} = \begin{pmatrix} \frac{3}{2}e^{-t} & -\frac{1}{2}e^{-t} \\ -\frac{1}{2}e^t & \frac{1}{2}e^t \end{pmatrix}$$

so that

$$\mathbf{U} = \int \boldsymbol{\Phi}^{-1}\mathbf{F}\,dt = \int \begin{pmatrix} -2te^{-t} \\ 2te^t \end{pmatrix} dt = \begin{pmatrix} 2te^{-t} + 2e^{-t} \\ 2te^t - 2e^t \end{pmatrix}$$

and

$$\mathbf{X}_p = \boldsymbol{\Phi}\mathbf{U} = \begin{pmatrix} 4 \\ 8 \end{pmatrix} t + \begin{pmatrix} 0 \\ -4 \end{pmatrix}.$$

15. From

$$\mathbf{X}' = \begin{pmatrix} 0 & 2 \\ -1 & 3 \end{pmatrix} \mathbf{X} + \begin{pmatrix} 1 \\ -1 \end{pmatrix} e^t$$

we obtain

$$\mathbf{X}_c = c_1 \begin{pmatrix} 2 \\ 1 \end{pmatrix} e^t + c_2 \begin{pmatrix} 1 \\ 1 \end{pmatrix} e^{2t}.$$

Then

$$\boldsymbol{\Phi} = \begin{pmatrix} 2e^t & e^{2t} \\ e^t & e^{2t} \end{pmatrix} \quad \text{and} \quad \boldsymbol{\Phi}^{-1} = \begin{pmatrix} e^{-t} & -e^{-t} \\ -e^{-2t} & 2e^{-2t} \end{pmatrix}$$

so that

$$\mathbf{U} = \int \boldsymbol{\Phi}^{-1}\mathbf{F}\,dt = \int \begin{pmatrix} 2 \\ -3e^{-t} \end{pmatrix} dt = \begin{pmatrix} 2t \\ 3e^{-t} \end{pmatrix}$$

and

$$\mathbf{X}_p = \boldsymbol{\Phi}\mathbf{U} = \begin{pmatrix} 4 \\ 2 \end{pmatrix} te^t + \begin{pmatrix} 3 \\ 3 \end{pmatrix} e^t.$$

18. From

$$\mathbf{X}' = \begin{pmatrix} 1 & 8 \\ 1 & -1 \end{pmatrix} \mathbf{X} + \begin{pmatrix} e^{-t} \\ te^{t} \end{pmatrix}$$

we obtain

$$\mathbf{X}_c = c_1 \begin{pmatrix} 4 \\ 1 \end{pmatrix} e^{3t} + c_2 \begin{pmatrix} -2 \\ 1 \end{pmatrix} e^{-3t}.$$

Then

$$\mathbf{\Phi} = \begin{pmatrix} 4e^{3t} & -2e^{3t} \\ e^{3t} & e^{-3t} \end{pmatrix} \quad \text{and} \quad \mathbf{\Phi}^{-1} = \begin{pmatrix} \frac{1}{6}e^{-3t} & \frac{1}{3}e^{-3t} \\ -\frac{1}{6}e^{3t} & \frac{2}{3}e^{3t} \end{pmatrix}$$

so that

$$\mathbf{U} = \int \mathbf{\Phi}^{-1}\mathbf{F}\,dt = \int \begin{pmatrix} \frac{1}{6}e^{-4t} + \frac{1}{3}te^{-2t} \\ -\frac{1}{6}e^{2t} + \frac{2}{3}te^{4t} \end{pmatrix} dt = \begin{pmatrix} -\frac{1}{24}e^{-4t} - \frac{1}{6}te^{-2t} - \frac{1}{12}e^{-2t} \\ -\frac{1}{12}e^{2t} + \frac{1}{6}te^{4t} - \frac{1}{24}e^{4t} \end{pmatrix}$$

and

$$\mathbf{X}_p = \mathbf{\Phi}\mathbf{U} = \begin{pmatrix} -te^{t} - \frac{1}{4}e^{t} \\ -\frac{1}{8}e^{-t} - \frac{1}{8}e^{t} \end{pmatrix}.$$

21. From

$$\mathbf{X}' = \begin{pmatrix} 0 & -1 \\ 1 & 0 \end{pmatrix} \mathbf{X} + \begin{pmatrix} \sec t \\ 0 \end{pmatrix}$$

we obtain

$$\mathbf{X}_c = c_1 \begin{pmatrix} \cos t \\ \sin t \end{pmatrix} + c_2 \begin{pmatrix} \sin t \\ -\cos t \end{pmatrix}.$$

Then

$$\mathbf{\Phi} = \begin{pmatrix} \cos t & \sin t \\ \sin t & -\cos t \end{pmatrix} \quad \text{and} \quad \mathbf{\Phi}^{-1} = \begin{pmatrix} \cos t & \sin t \\ \sin t & -\cos t \end{pmatrix}$$

so that

$$\mathbf{U} = \int \mathbf{\Phi}^{-1}\mathbf{F}\,dt = \int \begin{pmatrix} 1 \\ \tan t \end{pmatrix} dt = \begin{pmatrix} t \\ -\ln|\cos t| \end{pmatrix}$$

and

$$\mathbf{X}_p = \mathbf{\Phi}\mathbf{U} = \begin{pmatrix} t\cos t - \sin t \ln|\cos t| \\ t\sin t + \cos t \ln|\cos t| \end{pmatrix}.$$

24. From

$$\mathbf{X}' = \begin{pmatrix} 2 & -2 \\ 8 & -6 \end{pmatrix} \mathbf{X} + \begin{pmatrix} 1 \\ 3 \end{pmatrix} \frac{1}{t} e^{-2t}$$

we obtain

$$\mathbf{X}_c = c_1 \begin{pmatrix} 1 \\ 2 \end{pmatrix} e^{-2t} + c_2 \left[\begin{pmatrix} 1 \\ 2 \end{pmatrix} t e^{-2t} + \begin{pmatrix} 1/2 \\ 1/2 \end{pmatrix} e^{-2t} \right].$$

Then

$$\mathbf{\Phi} = \begin{pmatrix} 1 & t + \frac{1}{2} \\ 2 & 2t + \frac{1}{2} \end{pmatrix} e^{-2t} \quad \text{and} \quad \mathbf{\Phi}^{-1} = \begin{pmatrix} -4t - 1 & 2t + 1 \\ 4 & -2 \end{pmatrix} e^{2t}$$

so that

$$\mathbf{U} = \int \mathbf{\Phi}^{-1} \mathbf{F} \, dt = \int \begin{pmatrix} 2 + 2/t \\ -2/t \end{pmatrix} dt = \begin{pmatrix} 2t + 2\ln t \\ -2\ln t \end{pmatrix}$$

and

$$\mathbf{X}_p = \mathbf{\Phi} \mathbf{U} = \begin{pmatrix} 2t + \ln t - 2t\ln t \\ 4t + 3\ln t - 4t\ln t \end{pmatrix} e^{-2t}.$$

27. From

$$\mathbf{X}' = \begin{pmatrix} 1 & 2 \\ -1/2 & 1 \end{pmatrix} \mathbf{X} + \begin{pmatrix} \csc t \\ \sec t \end{pmatrix} e^t$$

we obtain

$$\mathbf{X}_c = c_1 \begin{pmatrix} 2\sin t \\ \cos t \end{pmatrix} e^t + c_2 \begin{pmatrix} 2\cos t \\ -\sin t \end{pmatrix} e^t.$$

Then

$$\mathbf{\Phi} = \begin{pmatrix} 2\sin t & 2\cos t \\ \cos t & -\sin t \end{pmatrix} e^t \quad \text{and} \quad \mathbf{\Phi}^{-1} = \begin{pmatrix} \frac{1}{2}\sin t & \cos t \\ \frac{1}{2}\cos t & -\sin t \end{pmatrix} e^{-t}$$

so that

$$\mathbf{U} = \int \mathbf{\Phi}^{-1} \mathbf{F} \, dt = \int \begin{pmatrix} \frac{3}{2} \\ \frac{1}{2}\cot t - \tan t \end{pmatrix} dt = \begin{pmatrix} \frac{3}{2}t \\ \frac{1}{2}\ln|\sin t| + \ln|\cos t| \end{pmatrix}$$

and

$$\mathbf{X}_p = \mathbf{\Phi} \mathbf{U} = \begin{pmatrix} 3\sin t \\ \frac{3}{2}\cos t \end{pmatrix} t e^t + \begin{pmatrix} \cos t \\ -\frac{1}{2}\sin t \end{pmatrix} e^t \ln|\sin t| + \begin{pmatrix} 2\cos t \\ -\sin t \end{pmatrix} e^t \ln|\cos t|.$$

30. From

$$\mathbf{X}' = \begin{pmatrix} 3 & -1 & -1 \\ 1 & 1 & -1 \\ 1 & -1 & 1 \end{pmatrix} \mathbf{X} + \begin{pmatrix} 0 \\ t \\ 2e^t \end{pmatrix}$$

we obtain

$$\mathbf{X}_c = c_1 \begin{pmatrix} 1 \\ 1 \\ 1 \end{pmatrix} e^t + c_2 \begin{pmatrix} 1 \\ 1 \\ 0 \end{pmatrix} e^{2t} + c_3 \begin{pmatrix} 1 \\ 0 \\ 1 \end{pmatrix} e^{2t}.$$

Then

$$\Phi = \begin{pmatrix} e^t & e^{2t} & e^{2t} \\ e^t & e^{2t} & 0 \\ e^t & 0 & e^{2t} \end{pmatrix} \quad \text{and} \quad \Phi^{-1} = \begin{pmatrix} -e^{-t} & e^{-t} & e^{-t} \\ e^{-2t} & 0 & -e^{-2t} \\ e^{-2t} & -e^{-2t} & 0 \end{pmatrix}$$

so that

$$\mathbf{U} = \int \Phi^{-1}\mathbf{F}\, dt = \int \begin{pmatrix} te^{-t} + 2 \\ -2e^{-t} \\ -te^{-2t} \end{pmatrix} dt = \begin{pmatrix} -te^{-t} - e^{-t} + 2t \\ 2e^{-t} \\ \frac{1}{2}te^{-2t} + \frac{1}{4}e^{-2t} \end{pmatrix}$$

and

$$\mathbf{X}_p = \Phi\mathbf{U} = \begin{pmatrix} -1/2 \\ -1 \\ -1/2 \end{pmatrix} t + \begin{pmatrix} -3/4 \\ -1 \\ -3/4 \end{pmatrix} + \begin{pmatrix} 2 \\ 2 \\ 0 \end{pmatrix} e^t + \begin{pmatrix} 2 \\ 2 \\ 2 \end{pmatrix} te^t.$$

33. Let $\mathbf{I} = \begin{pmatrix} i_1 \\ i_2 \end{pmatrix}$ so that

$$\mathbf{I}' = \begin{pmatrix} -11 & 3 \\ 3 & -3 \end{pmatrix} \mathbf{I} + \begin{pmatrix} 100\sin t \\ 0 \end{pmatrix}$$

and

$$\mathbf{I}_c = c_1 \begin{pmatrix} 1 \\ 3 \end{pmatrix} e^{-2t} + c_2 \begin{pmatrix} 3 \\ -1 \end{pmatrix} e^{-12t}.$$

Then

$$\Phi = \begin{pmatrix} e^{-2t} & 3e^{-12t} \\ 3e^{-2t} & -e^{-12t} \end{pmatrix}, \quad \Phi^{-1} = \begin{pmatrix} \frac{1}{10}e^{2t} & \frac{3}{10}e^{2t} \\ \frac{3}{10}e^{12t} & -\frac{1}{10}e^{12t} \end{pmatrix},$$

$$\mathbf{U} = \int \Phi^{-1}\mathbf{F}\, dt = \int \begin{pmatrix} 10e^{2t}\sin t \\ 30e^{12t}\sin t \end{pmatrix} dt = \begin{pmatrix} 2e^{2t}(2\sin t - \cos t) \\ \frac{6}{29}e^{12t}(12\sin t - \cos t) \end{pmatrix},$$

and

$$\mathbf{I}_p = \Phi\mathbf{U} = \begin{pmatrix} \frac{332}{29}\sin t - \frac{76}{29}\cos t \\ \frac{276}{29}\sin t - \frac{168}{29}\cos t \end{pmatrix}$$

so that

$$\mathbf{I} = c_1 \begin{pmatrix} 1 \\ 3 \end{pmatrix} e^{-2t} + c_2 \begin{pmatrix} 3 \\ -1 \end{pmatrix} e^{-12t} + \mathbf{I}_p.$$

If $\mathbf{I}(0) = \begin{pmatrix} 0 \\ 0 \end{pmatrix}$ then $c_1 = 2$ and $c_2 = \frac{6}{29}$.

8.4 | Matrix Exponential

The terminology and concepts listed below provide an outline of the main ideas encountered in this section. These can be useful when preparing for a quiz or test.

Terminology and Concepts

- infinite series representation of the matrix exponential $e^{\mathbf{A}t}$
- the matrix exponential is a fundamental matrix

The basic skills listed below summarize the more mechanical types of problems encountered in the exercise set for this section.

Basic Skills

- compute $e^{\mathbf{A}t}$ for a given matrix $\mathbf{A}$
- use the matrix exponential to solve a nonhomogeneous linear system

Use of Computers   To compute the matrix exponential for a square matrix $\mathbf{A}t$ use

MatrixExp[At]	(*Mathematica*)
with(linalg):	(*Maple*)
exponential(A,t);	
expm(At)	(MATLAB)

3. For

$$\mathbf{A} = \begin{pmatrix} 1 & 1 & 1 \\ 1 & 1 & 1 \\ -2 & -2 & -2 \end{pmatrix}$$

we have

$$\mathbf{A}^2 = \begin{pmatrix} 1 & 1 & 1 \\ 1 & 1 & 1 \\ -2 & -2 & -2 \end{pmatrix}\begin{pmatrix} 1 & 1 & 1 \\ 1 & 1 & 1 \\ -2 & -2 & -2 \end{pmatrix} = \begin{pmatrix} 0 & 0 & 0 \\ 0 & 0 & 0 \\ 0 & 0 & 0 \end{pmatrix}.$$

Thus, $\mathbf{A}^3 = \mathbf{A}^4 = \mathbf{A}^5 = \cdots = \mathbf{0}$ and

$$e^{\mathbf{A}t} = \mathbf{I} + \mathbf{A}t = \begin{pmatrix} 1 & 0 & 0 \\ 0 & 1 & 0 \\ 0 & 0 & 1 \end{pmatrix} + \begin{pmatrix} t & t & t \\ t & t & t \\ -2t & -2t & -2t \end{pmatrix} = \begin{pmatrix} t+1 & t & t \\ t & t+1 & t \\ -2t & -2t & -2t+1 \end{pmatrix}.$$

6. Using the result of Problem 2,

$$\mathbf{X} = \begin{pmatrix} \cosh t & \sinh t \\ \sinh t & \cosh t \end{pmatrix} \begin{pmatrix} c_1 \\ c_2 \end{pmatrix} = c_1 \begin{pmatrix} \cosh t \\ \sinh t \end{pmatrix} + c_2 \begin{pmatrix} \sinh t \\ \cosh t \end{pmatrix}.$$

9. To solve

$$\mathbf{X}' = \begin{pmatrix} 1 & 0 \\ 0 & 2 \end{pmatrix} \mathbf{X} + \begin{pmatrix} 3 \\ -1 \end{pmatrix}$$

we identify $t_0 = 0$, $\mathbf{F}(t) = \begin{pmatrix} 3 \\ -1 \end{pmatrix}$, and use the results of Problem 1 and equation (5) in the text.

$$\mathbf{X}(t) = e^{\mathbf{A}t}\mathbf{C} + e^{\mathbf{A}t} \int_{t_0}^{t} e^{-\mathbf{A}s}\mathbf{F}(s)\, ds$$

$$= \begin{pmatrix} e^t & 0 \\ 0 & e^{2t} \end{pmatrix} \begin{pmatrix} c_1 \\ c_2 \end{pmatrix} + \begin{pmatrix} e^t & 0 \\ 0 & e^{2t} \end{pmatrix} \int_0^t \begin{pmatrix} e^{-s} & 0 \\ 0 & e^{-2s} \end{pmatrix} \begin{pmatrix} 3 \\ -1 \end{pmatrix} ds$$

$$= \begin{pmatrix} c_1 e^t \\ c_2 e^{2t} \end{pmatrix} + \begin{pmatrix} e^t & 0 \\ 0 & e^{2t} \end{pmatrix} \int_0^t \begin{pmatrix} 3e^{-s} \\ -e^{-2s} \end{pmatrix} ds$$

$$= \begin{pmatrix} c_1 e^t \\ c_2 e^{2t} \end{pmatrix} + \begin{pmatrix} e^t & 0 \\ 0 & e^{2t} \end{pmatrix} \begin{pmatrix} -3e^{-s} \\ \frac{1}{2}e^{-2s} \end{pmatrix} \Big|_0^t$$

$$= \begin{pmatrix} c_1 e^t \\ c_2 e^{2t} \end{pmatrix} + \begin{pmatrix} e^t & 0 \\ 0 & e^{2t} \end{pmatrix} \begin{pmatrix} -3e^{-t} + 3 \\ \frac{1}{2}e^{-2t} - \frac{1}{2} \end{pmatrix}$$

$$= \begin{pmatrix} c_1 e^t \\ c_2 e^{2t} \end{pmatrix} + \begin{pmatrix} -3 + 3e^t \\ \frac{1}{2} - \frac{1}{2}e^{2t} \end{pmatrix} = c_3 \begin{pmatrix} 1 \\ 0 \end{pmatrix} e^t + c_4 \begin{pmatrix} 0 \\ 1 \end{pmatrix} e^{2t} + \begin{pmatrix} -3 \\ \frac{1}{2} \end{pmatrix}.$$

12. To solve

$$\mathbf{X}' = \begin{pmatrix} 0 & 1 \\ 1 & 0 \end{pmatrix} \mathbf{X} + \begin{pmatrix} \cosh t \\ \sinh t \end{pmatrix}$$

we identify $t_0 = 0$, $\mathbf{F}(t) = \begin{pmatrix} \cosh t \\ \sinh t \end{pmatrix}$, and use the results of Problem 2 and equation (5) in the text.

$$\mathbf{X}(t) = e^{\mathbf{A}t}\mathbf{C} + e^{\mathbf{A}t} \int_{t_0}^{t} e^{-\mathbf{A}s}\mathbf{F}(s)\, ds$$

$$= \begin{pmatrix} \cosh t & \sinh t \\ \sinh t & \cosh t \end{pmatrix} \begin{pmatrix} c_1 \\ c_2 \end{pmatrix} + \begin{pmatrix} \cosh t & \sinh t \\ \sinh t & \cosh t \end{pmatrix} \int_0^t \begin{pmatrix} \cosh s & -\sinh s \\ -\sinh s & \cosh s \end{pmatrix} \begin{pmatrix} \cosh s \\ \sinh s \end{pmatrix} ds$$

$$= \begin{pmatrix} c_1 \cosh t + c_2 \sinh t \\ c_1 \sinh t + c_2 \cosh t \end{pmatrix} + \begin{pmatrix} \cosh t & \sinh t \\ \sinh t & \cosh t \end{pmatrix} \int_0^t \begin{pmatrix} 1 \\ 0 \end{pmatrix} ds$$

$$= \begin{pmatrix} c_1 \cosh t + c_2 \sinh t \\ c_1 \sinh t + c_2 \cosh t \end{pmatrix} + \begin{pmatrix} \cosh t & \sinh t \\ \sinh t & \cosh t \end{pmatrix} \begin{pmatrix} s \\ 0 \end{pmatrix} \Big|_0^t$$

$$= \begin{pmatrix} c_1 \cosh t + c_2 \sinh t \\ c_1 \sinh t + c_2 \cosh t \end{pmatrix} + \begin{pmatrix} \cosh t & \sinh t \\ \sinh t & \cosh t \end{pmatrix} \begin{pmatrix} t \\ 0 \end{pmatrix}$$

$$= \begin{pmatrix} c_1 \cosh t + c_2 \sinh t \\ c_1 \sinh t + c_2 \cosh t \end{pmatrix} + \begin{pmatrix} t \cosh t \\ t \sinh t \end{pmatrix} = c_1 \begin{pmatrix} \cosh t \\ \sinh t \end{pmatrix} + c_2 \begin{pmatrix} \sinh t \\ \cosh t \end{pmatrix} + t \begin{pmatrix} \cosh t \\ \sinh t \end{pmatrix}.$$

15. From $s\mathbf{I} - \mathbf{A} = \begin{pmatrix} s & 4 & 3 \\ 4 & s+4 \end{pmatrix}$ we find

$$(s\mathbf{I} - \mathbf{A})^{-1} = \begin{pmatrix} \dfrac{3/2}{s-2} - \dfrac{1/2}{s+2} & \dfrac{3/4}{s-2} - \dfrac{3/4}{s+2} \\ \dfrac{-1}{s-2} + \dfrac{1}{s+2} & \dfrac{-1/2}{s-2} + \dfrac{3/2}{s+2} \end{pmatrix}$$

and

$$e^{\mathbf{A}t} = \begin{pmatrix} \frac{3}{2}e^{2t} - \frac{1}{2}e^{-2t} & \frac{3}{4}e^{2t} - \frac{3}{4}e^{-2t} \\ -e^{2t} + e^{-2t} & -\frac{1}{2}e^{2t} + \frac{3}{2}e^{-2t} \end{pmatrix}.$$

The general solution of the system is then

$$\mathbf{X} = e^{\mathbf{A}t}\mathbf{C} = \begin{pmatrix} \frac{3}{2}e^{2t} - \frac{1}{2}e^{-2t} & \frac{3}{4}e^{2t} - \frac{3}{4}e^{-2t} \\ -e^{2t} + e^{-2t} & -\frac{1}{2}e^{2t} + \frac{3}{2}e^{-2t} \end{pmatrix} \begin{pmatrix} c_1 \\ c_2 \end{pmatrix}$$

$$= c_1 \begin{pmatrix} 3/2 \\ -1 \end{pmatrix} e^{2t} + c_1 \begin{pmatrix} -1/2 \\ 1 \end{pmatrix} e^{-2t} + c_2 \begin{pmatrix} 3/4 \\ -1/2 \end{pmatrix} e^{2t} + c_2 \begin{pmatrix} -3/4 \\ 3/2 \end{pmatrix} e^{-2t}$$

$$= \left(\frac{1}{2}c_1 + \frac{1}{4}c_2\right) \begin{pmatrix} 3 \\ -2 \end{pmatrix} e^{2t} + \left(-\frac{1}{2}c_1 - \frac{3}{4}c_2\right) \begin{pmatrix} 1 \\ -2 \end{pmatrix} e^{-2t}$$

$$= c_3 \begin{pmatrix} 3 \\ -2 \end{pmatrix} e^{2t} + c_4 \begin{pmatrix} 1 \\ -2 \end{pmatrix} e^{-2t}.$$

18. From $s\mathbf{I} - \mathbf{A} = \begin{pmatrix} s & -1 \\ 2 & s+2 \end{pmatrix}$ we find

$$(s\mathbf{I} - \mathbf{A})^{-1} = \begin{pmatrix} \dfrac{s+1+1}{(s+1)^2+1} & \dfrac{1}{(s+1)^2+1} \\[3mm] \dfrac{-2}{(s+1)^2+1} & \dfrac{s+1-1}{(s+1)^2+1} \end{pmatrix}$$

and

$$e^{\mathbf{A}t} = \begin{pmatrix} e^{-t}\cos t + e^{-t}\sin t & e^{-t}\sin t \\ -2e^{-t}\sin t & e^{-t}\cos t - e^{-t}\sin t \end{pmatrix}.$$

The general solution of the system is then

$$\mathbf{X} = e^{\mathbf{A}t}\mathbf{C} = \begin{pmatrix} e^{-t}\cos t + e^{-t}\sin t & e^{-t}\sin t \\ -2e^{-t}\sin t & e^{-t}\cos t - e^{-t}\sin t \end{pmatrix}\begin{pmatrix} c_1 \\ c_2 \end{pmatrix}$$

$$= c_1\begin{pmatrix} 1 \\ 0 \end{pmatrix}e^{-t}\cos t + c_1\begin{pmatrix} 1 \\ -2 \end{pmatrix}e^{-t}\sin t + c_2\begin{pmatrix} 0 \\ 1 \end{pmatrix}e^{-t}\cos t + c_2\begin{pmatrix} 1 \\ -1 \end{pmatrix}e^{-t}\sin t$$

$$= c_1\begin{pmatrix} \cos t + \sin t \\ -2\sin t \end{pmatrix}e^{-t} + c_2\begin{pmatrix} \sin t \\ \cos t - \sin t \end{pmatrix}e^{-t}.$$

21. From equation (3) in the text

$$e^{t\mathbf{A}} = e^{t\mathbf{PDP}^{-1}} = \mathbf{I} + t(\mathbf{PDP}^{-1}) + \frac{1}{2!}t^2(\mathbf{PDP}^{-1})^2 + \frac{1}{3!}t^3(\mathbf{PDP}^{-1})^3 + \cdots$$

$$= \mathbf{P}\left[\mathbf{I} + t\mathbf{D} + \frac{1}{2!}(t\mathbf{D})^2 + \frac{1}{3!}(t\mathbf{D})^3 + \cdots\right]\mathbf{P}^{-1} = \mathbf{P}e^{t\mathbf{D}}\mathbf{P}^{-1}.$$

24. From Problems 20-22 and equation (1) in the text

$$\mathbf{X} = e^{t\mathbf{A}}\mathbf{C} = \mathbf{P}e^{t\mathbf{D}}\mathbf{P}^{-1}\mathbf{C}$$

$$= \begin{pmatrix} -e^t & e^{3t} \\ e^t & e^{3t} \end{pmatrix}\begin{pmatrix} e^t & 0 \\ 0 & e^{3t} \end{pmatrix}\begin{pmatrix} -\frac{1}{2}e^{-t} & \frac{1}{2}e^{-t} \\ \frac{1}{2}e^{3t} & \frac{1}{2}e^{-3t} \end{pmatrix}\begin{pmatrix} c_1 \\ c_2 \end{pmatrix}$$

$$= \begin{pmatrix} \frac{1}{2}e^t + \frac{1}{2}e^{9t} & -\frac{1}{2}e^t + \frac{1}{2}e^{3t} \\ -\frac{1}{2}e^t + \frac{1}{2}e^{9t} & \frac{1}{2}e^t + \frac{1}{2}e^{3t} \end{pmatrix}\begin{pmatrix} c_1 \\ c_2 \end{pmatrix}.$$

27. (a) The following commands can be used in *Mathematica*:

$A=\{\{4, 2\},\{3, 3\}\};$

$c=\{c1, c2\};$

$m=MatrixExp[A\ t];$

$sol=Expand[m.c]$

$Collect[sol, \{c1, c2\}]//MatrixForm$

The output gives

$$x(t) = c_1 \left(\frac{2}{5}e^t + \frac{3}{5}e^{6t} \right) + c_2 \left(-\frac{2}{5}e^t + \frac{2}{5}e^{6t} \right)$$

$$y(t) = c_1 \left(-\frac{3}{5}e^t + \frac{3}{5}e^{6t} \right) + c_2 \left(\frac{3}{5}e^t + \frac{2}{5}e^{6t} \right).$$

The eigenvalues are 1 and 6 with corresponding eigenvectors

$$\begin{pmatrix} -2 \\ 3 \end{pmatrix} \quad \text{and} \quad \begin{pmatrix} 1 \\ 1 \end{pmatrix},$$

so the solution of the system is

$$\mathbf{X}(t) = b_1 \begin{pmatrix} -2 \\ 3 \end{pmatrix} e^t + b_2 \begin{pmatrix} 1 \\ 1 \end{pmatrix} e^{6t}$$

or

$$x(t) = -2b_1 e^t + b_2 e^{6t}$$

$$y(t) = 3b_1 e^t + b_2 e^{6t}.$$

If we replace b_1 with $-\frac{1}{5}c_1 + \frac{1}{5}c_2$ and b_2 with $\frac{3}{5}c_1 + \frac{2}{5}c_2$, we obtain the solution found using the matrix exponential.

(b) $x(t) = c_1 e^{-2t} \cos t - (c_1 + c_2)e^{-2t} \sin t$

$y(t) = c_2 e^{-2t} \cos t + (2c_1 + c_2)e^{-2t} \sin t$

8.R   Chapter 8 in Review

3. Since

$$\begin{pmatrix} 4 & 6 & 6 \\ 1 & 3 & 2 \\ -1 & -4 & -3 \end{pmatrix} \begin{pmatrix} 3 \\ 1 \\ -1 \end{pmatrix} = \begin{pmatrix} 12 \\ 4 \\ -4 \end{pmatrix} = 4 \begin{pmatrix} 3 \\ 1 \\ -1 \end{pmatrix},$$

we see that $\lambda = 4$ is an eigenvalue with eigenvector $\mathbf{K}_3$. The corresponding solution is $\mathbf{X}_3 = \mathbf{K}_3 e^{4t}$.

6. We have $\det(\mathbf{A} - \lambda\mathbf{I}) = (\lambda + 6)(\lambda + 2) = 0$ so that

$$\mathbf{X} = c_1 \begin{pmatrix} 1 \\ -1 \end{pmatrix} e^{-6t} + c_2 \begin{pmatrix} 1 \\ 1 \end{pmatrix} e^{-2t}.$$

9. We have $\det(\mathbf{A} - \lambda\mathbf{I}) = -(\lambda - 2)(\lambda - 4)(\lambda + 3) = 0$ so that

$$\mathbf{X} = c_1 \begin{pmatrix} -2 \\ 3 \\ 1 \end{pmatrix} e^{2t} + c_2 \begin{pmatrix} 0 \\ 1 \\ 1 \end{pmatrix} e^{4t} + c_3 \begin{pmatrix} 7 \\ 12 \\ -16 \end{pmatrix} e^{-3t}.$$

12. We have

$$\mathbf{X}_c = c_1 \begin{pmatrix} 2\cos t \\ -\sin t \end{pmatrix} e^t + c_2 \begin{pmatrix} 2\sin t \\ \cos t \end{pmatrix} e^t.$$

Then

$$\mathbf{\Phi} = \begin{pmatrix} 2\cos t & 2\sin t \\ -\sin t & \cos t \end{pmatrix} e^t, \quad \mathbf{\Phi}^{-1} = \begin{pmatrix} \frac{1}{2}\cos t & -\sin t \\ \frac{1}{2}\sin t & \cos t \end{pmatrix} e^{-t},$$

and

$$\mathbf{U} = \int \mathbf{\Phi}^{-1}\mathbf{F}\, dt = \int \begin{pmatrix} \cos t - \sec t \\ \sin t \end{pmatrix} dt = \begin{pmatrix} \sin t - \ln|\sec t + \tan t| \\ -\cos t \end{pmatrix},$$

so that

$$\mathbf{X}_p = \mathbf{\Phi}\mathbf{U} = \begin{pmatrix} -2\cos t \ln|\sec t + \tan t| \\ -1 + \sin t \ln|\sec t + \tan t| \end{pmatrix} e^t.$$

15. (a) Letting

$$\mathbf{K} = \begin{pmatrix} k_1 \\ k_2 \\ k_3 \end{pmatrix}$$

we note that $(\mathbf{A} - 2\mathbf{I})\mathbf{K} = \mathbf{0}$ implies that $3k_1 + 3k_2 + 3k_3 = 0$, so $k_1 = -(k_2 + k_3)$. Choosing $k_2 = 0$, $k_3 = 1$ and then $k_2 = 1$, $k_3 = 0$ we get

$$\mathbf{K}_1 = \begin{pmatrix} -1 \\ 0 \\ 1 \end{pmatrix} \quad \text{and} \quad \mathbf{K}_2 = \begin{pmatrix} -1 \\ 1 \\ 0 \end{pmatrix},$$

respectively. Thus,

$$\mathbf{X}_1 = \begin{pmatrix} -1 \\ 0 \\ 1 \end{pmatrix} e^{2t} \quad \text{and} \quad \mathbf{X}_2 = \begin{pmatrix} -1 \\ 1 \\ 0 \end{pmatrix} e^{2t}$$

are two solutions.

(b) From $\det(\mathbf{A} - \lambda\mathbf{I}) = \lambda^2(3 - \lambda) = 0$ we see that $\lambda_1 = 3$, and 0 is an eigenvalue of multiplicity two. Letting

$$\mathbf{K} = \begin{pmatrix} k_1 \\ k_2 \\ k_3 \end{pmatrix},$$

as in part **(a)**, we note that $(\mathbf{A} - 0\mathbf{I})\mathbf{K} = \mathbf{AK} = \mathbf{0}$ implies that $k_1 + k_2 + k_3 = 0$, so $k_1 = -(k_2 + k_3)$. Choosing $k_2 = 0$. $k_3 = 1$, and then $k_2 = 1$, $k_3 = 0$ we get

$$\mathbf{K}_2 = \begin{pmatrix} -1 \\ 0 \\ 1 \end{pmatrix} \quad \text{and} \quad \mathbf{K}_3 = \begin{pmatrix} -1 \\ 1 \\ 0 \end{pmatrix},$$

respectively. Since the eigenvector corresponding to $\lambda_1 = 3$ is

$$\mathbf{K}_1 = \begin{pmatrix} 1 \\ 1 \\ 1 \end{pmatrix},$$

the general solution of the system is

$$\mathbf{X} = c_1 \begin{pmatrix} 1 \\ 1 \\ 1 \end{pmatrix} e^{3t} + c_2 \begin{pmatrix} -1 \\ 0 \\ 1 \end{pmatrix} + c_3 \begin{pmatrix} -1 \\ 1 \\ 0 \end{pmatrix}.$$

9 Numerical Solutions of Ordinary Differential Equations

9.1 Euler Methods and Error Analysis

The terminology and concepts listed below provide an outline of the main ideas encountered in this section. These can be useful when preparing for a quiz or test.

Terminology and Concepts

- error analysis for numerical solutions of ordinary DEs

- round-off error

- local and global truncation errors

- bound for local truncation error

- Euler's method for approximating the solution of a first-order IVP

- improved Euler's method for approximating the solution of a first-order IVP

- a predictor-corrector method

The basic skills listed below summarize the more mechanical types of problems encountered in the exercise set for this section.

Basic Skills

- use the improved Euler's method to approximate the solution of a first-order IVP

- find a bound on the local truncation error when Euler's method or the improved Euler's method is used to approximate the solution of an IVP

233

3.

$h=0.1$

x_n	y_n
0.00	0.0000
0.10	0.1005
0.20	0.2030
0.30	0.3098
0.40	0.4234
0.50	0.5470

$h=0.05$

x_n	y_n
0.00	0.0000
0.05	0.0501
0.10	0.1004
0.15	0.1512
0.20	0.2028
0.25	0.2554
0.30	0.3095
0.35	0.3652
0.40	0.4230
0.45	0.4832
0.50	0.5465

6.

$h=0.1$

x_n	y_n
0.00	0.0000
0.10	0.0050
0.20	0.0200
0.30	0.0451
0.40	0.0805
0.50	0.1266

$h=0.05$

x_n	y_n
0.00	0.0000
0.05	0.0013
0.10	0.0050
0.15	0.0113
0.20	0.0200
0.25	0.0313
0.30	0.0451
0.35	0.0615
0.40	0.0805
0.45	0.1022
0.50	0.1266

9.

$h=0.1$

x_n	y_n
1.00	1.0000
1.10	1.0095
1.20	1.0404
1.30	1.0967
1.40	1.1866
1.50	1.3260

$h=0.05$

x_n	y_n
1.00	1.0000
1.05	1.0024
1.10	1.0100
1.15	1.0228
1.20	1.0414
1.25	1.0663
1.30	1.0984
1.35	1.1389
1.40	1.1895
1.45	1.2526
1.50	1.3315

12. (a)

(b)

x_n	Euler	Imp. Euler
1.00	1.0000	1.0000
1.10	1.2000	1.2469
1.20	1.4938	1.6430
1.30	1.9711	2.4042
1.40	2.9060	4.5085

15. (a) Using Euler's method we obtain $y(0.1) \approx y_1 = 0.8$.

(b) Using $y'' = 5e^{-2x}$ we see that the local truncation error is

$$5e^{-2c}\frac{(0.1)^2}{2} = 0.025e^{-2c}.$$

Since e^{-2x} is a decreasing function, $e^{-2c} \leq e^0 = 1$ for $0 \leq c \leq 0.1$. Thus an upper bound for the local truncation error is $0.025(1) = 0.025$.

(c) Since $y(0.1) = 0.8234$, the actual error is $y(0.1) - y_1 = 0.0234$, which is less than 0.025.

(d) Using Euler's method with $h = 0.05$ we obtain $y(0.1) \approx y_2 = 0.8125$.

(e) The error in (d) is $0.8234 - 0.8125 = 0.0109$. With global truncation error $O(h)$, when the step size is halved we expect the error for $h = 0.05$ to be one-half the error when $h = 0.1$. Comparing 0.0109 with 0.0234 we see that this is the case.

18. (a) Using $y''' = -114e^{-3(x-1)}$ we see that the local truncation error is

$$\left| y'''(c) \frac{h^3}{6} \right| = 114e^{-3(x-1)} \frac{h^3}{6} = 19h^3 e^{-3(c-1)}.$$

(b) Since $e^{-3(x-1)}$ is a decreasing function for $1 \le x \le 1.5$, $e^{-3(c-1)} \le e^{-3(1-1)} = 1$ for $1 \le c \le 1.5$ and

$$\left| y'''(c) \frac{h^3}{6} \right| \le 19(0.1)^3(1) = 0.019.$$

(c) Using the improved Euler's method with $h = 0.1$ we obtain $y(1.5) \approx 2.080108$. With $h = 0.05$ we obtain $y(1.5) \approx 2.059166$.

(d) Since $y(1.5) = 2.053216$, the error for $h = 0.1$ is $E_{0.1} = 0.026892$, while the error for $h = 0.05$ is $E_{0.05} = 0.005950$. With global truncation error $O(h^2)$ we expect $E_{0.1}/E_{0.05} \approx 4$. We actually have $E_{0.1}/E_{0.05} = 4.52$.

21. Because y_{n+1}^* depends on y_n and is used to determine y_{n+1}, all of the y_n^* cannot be computed at one time independently of the corresponding y_n values. For example, the computation of y_4^* involves the value of y_3.

9.2 │ Runge-Kutta Methods

The terminology and concepts listed below provide an outline of the main ideas encountered in this section. These can be useful when preparing for a quiz or test.

Terminology and Concepts

- Runge-Kutta methods

- fourth-order Runge-Kutta method (RK4 method)

- truncation errors for the RK4 method

The basic skills listed below summarize the more mechanical types of problems encountered in the exercise set for this section.

Basic Skills

- use the RK4 method to approximate the solution of a first-order initial-value problem

3.

x_n	y_n
1.00	5.0000
1.10	3.9724
1.20	3.2284
1.30	2.6945
1.40	2.3163
1.50	2.0533

6.

x_n	y_n
0.00	1.0000
0.10	1.1115
0.20	1.2530
0.30	1.4397
0.40	1.6961
0.50	2.0670

9.

x_n	y_n
0.00	0.5000
0.10	0.5213
0.20	0.5358
0.30	0.5443
0.40	0.5482
0.50	0.5493

12.

x_n	y_n
0.00	0.5000
0.10	0.5250
0.20	0.5498
0.30	0.5744
0.40	0.5987
0.50	0.6225

15. (a)

x_n	$h=0.05$	$h=0.1$
1.00	1.0000	1.0000
1.05	1.1112	
1.10	1.2511	1.2511
1.15	1.4348	
1.20	1.6934	1.6934
1.25	2.1047	
1.30	2.9560	2.9425
1.35	7.8981	
1.40	1.0608×10^{15}	903.0282

(b)

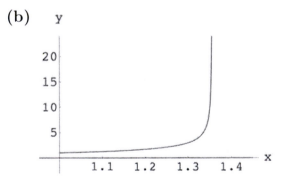

18. (a) Using $y^{(5)} = -1026e^{-3(x-1)}$ we see that the local truncation error is

$$\left| y^{(5)}(c) \frac{h^5}{120} \right| = 8.55 h^5 e^{-3(c-1)}.$$

(b) Since $e^{-3(x-1)}$ is a decreasing function for $1 \leq x \leq 1.5$, $e^{-3(c-1)} \leq e^{-3(1-1)} = 1$ for $1 \leq c \leq 1.5$ and

$$y^{(5)}(c) \frac{h^5}{120} \leq 8.55(0.1)^5(1) = 0.0000855.$$

(c) Using the RK4 method with $h = 0.1$ we obtain $y(1.5) \approx 2.053338827$. With $h = 0.05$ we obtain $y(1.5) \approx 2.053222989$.

21. (a) For $y' + y = 10 \sin 3x$ an integrating factor is e^x so that

$$\frac{d}{dx}[e^x y] = 10 e^x \sin 3x \implies e^x y = e^x \sin 3x - 3 e^x \cos 3x + c$$

$$\implies y = \sin 3x - 3 \cos 3x + c e^{-x}.$$

When $x = 0$, $y = 0$, so $0 = -3 + c$ and $c = 3$. The solution is

$$y = \sin 3x - 3 \cos 3x + 3 e^{-x}.$$

Using Newton's method we find that $x = 1.53235$ is the only positive root in $[0, 2]$.

(b) Using the RK4 method with $h = 0.1$ we obtain the table of values shown. These values are used to obtain an interpolating function in *Mathematica*. The graph of the interpolating function is shown. Using *Mathematica*'s root finding capability we see that the only positive root in $[0, 2]$ is $x = 1.53236$.

x_n	y_n	x_n	y_n
0.0	0.0000	1.0	4.2147
0.1	0.1440	1.1	3.8033
0.2	0.5448	1.2	3.1513
0.3	1.1409	1.3	2.3076
0.4	1.8559	1.4	1.3390
0.5	2.6049	1.5	0.3243
0.6	3.3019	1.6	-0.6530
0.7	3.8675	1.7	-1.5117
0.8	4.2356	1.8	-2.1809
0.9	4.3593	1.9	-2.6061
1.0	4.2147	2.0	-2.7539

9.3 | Multistep Methods

The terminology and concepts listed below provide an outline of the main ideas encountered in this section. These can be useful when preparing for a quiz or test.

Terminology and Concepts

- predictor-corrector methods
- Adams-Bashforth-Moulton method

The basic skills listed below summarize the more mechanical types of problems encountered in the exercise set for this section.

Basic Skills

- use the Adams-Bashforth-Moulton predictor-corrector method to approximate the solution of a first-order IVP

In the tables in this section "ABM" stands for Adams-Bashforth-Moulton.

3. The first predictor is $y_4^* = 0.73318477$.

x_n	y_n	
0.0	1.00000000	init. cond.
0.2	0.73280000	RK4
0.4	0.64608032	RK4
0.6	0.65851653	RK4
0.8	0.72319464	ABM

6. The first predictor for $h = 0.2$ is $y_4^* = 3.34828434$.

x_n	h=0.2		h=0.1	
0.0	1.00000000	init. cond.	1.00000000	init. cond.
0.1			1.21017082	RK4
0.2	1.44139950	RK4	1.44140511	RK4
0.3			1.69487942	RK4
0.4	1.97190167	RK4	1.97191536	ABM
0.5			2.27400341	ABM
0.6	2.60280694	RK4	2.60283209	ABM
0.7			2.96031780	ABM
0.8	3.34860927	ABM	3.34863769	ABM
0.9			3.77026548	ABM
1.0	4.22797875	ABM	4.22801028	ABM

The terminology and concepts listed below provide an outline of the main ideas encountered in this section. These can be useful when preparing for a quiz or test.

Terminology and Concepts

- express a second-order IVP as an IVP for a first-order system

- express a higher-order IVP as an IVP for a first-order system

- numerical solution of an IVP for a first-order system using Euler's method and the RK4 method

The basic skills listed below summarize the more mechanical types of problems encountered in the exercise set for this section.

Basic Skills

- solve an IVP for a first-order system consisting of two equations

3. The substitution $y' = u$ leads to the system

$$y' = u, \qquad u' = 4u - 4y.$$

Using formula (4) in the text with x corresponding to t, y corresponding to x, and u corresponding to y, we obtain the table shown.

x_n	h=0.2 y_n	h=0.2 u_n	h=0.1 y_n	h=0.1 u_n
0.0	-2.0000	1.0000	-2.0000	1.0000
0.1			-1.8321	2.4427
0.2	-1.4928	4.4731	-1.4919	4.4753

6. Using $h = 0.1$, the RK4 method for a system, and a numerical solver, we obtain

t_n	h=0.2 i_{1n}	h=0.2 i_{3n}
0.0	0.0000	0.0000
0.1	2.5000	3.7500
0.2	2.8125	5.7813
0.3	2.0703	7.4023
0.4	0.6104	9.1919
0.5	-1.5619	11.4877

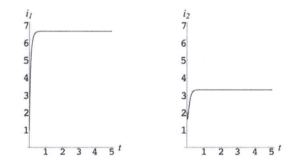

9.

t_n	h=0.2 x_n	h=0.2 y_n	h=0.1 x_n	h=0.1 y_n
0.0	-3.0000	5.0000	-3.0000	5.0000
0.1			-3.4790	4.6707
0.2	-3.9123	4.2857	-3.9123	4.2857

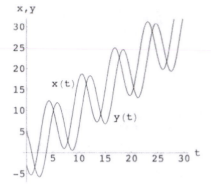

12. Solving for x' and y' we obtain the system

$$x' = \frac{1}{2}y - 3t^2 + 2t - 5$$

$$y' = -\frac{1}{2}y + 3t^2 + 2t + 5.$$

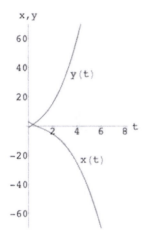

t_n	h=0.2 x_n	h=0.2 y_n	h=0.1 x_n	h=0.1 y_n
0.0	3.0000	-1.0000	3.0000	-1.0000
0.1			2.4727	-0.4527
0.2	1.9867	0.0933	1.9867	0.0933

9.5 | Second-Order Boundary-Value Problems

The terminology and concepts listed below provide an outline of the main ideas encountered in this section. These can be useful when preparing for a quiz or test.

Terminology and Concepts

- difference quotient

- finite difference

- forward difference

- backward difference

- central difference

- finite difference method

The basic skills listed below summarize the more mechanical types of problems encountered in the exercise set for this section.

Basic Skills

- use the finite difference method to approximate the solution of a second-order boundary-value problem

3. We identify $P(x) = 2$, $Q(x) = 1$, $f(x) = 5x$, and $h = (1-0)/5 = 0.2$. Then the finite difference equation is

$$1.2y_{i+1} - 1.96y_i + 0.8y_{i-1} = 0.04(5x_i).$$

The solution of the corresponding linear system gives

x	0.0	0.2	0.4	0.6	0.8	1.0
y	0.0000	-0.2259	-0.3356	-0.3308	-0.2167	0.0000

6. We identify $P(x) = 5$, $Q(x) = 0$, $f(x) = 4\sqrt{x}$, and $h = (2-1)/6 = 0.1667$. Then the finite difference equation is

$$1.4167y_{i+1} - 2y_i + 0.5833y_{i-1} = 0.2778(4\sqrt{x_i}).$$

The solution of the corresponding linear system gives

x	1.0000	1.1667	1.3333	1.5000	1.6667	1.8333	2.0000
y	1.0000	-0.5918	-1.1626	-1.3070	-1.2704	-1.1541	-1.0000

9. We identify $P(x) = 1 - x$, $Q(x) = x$, $f(x) = x$, and $h = (1-0)/10 = 0.1$. Then the finite difference equation is

$$[1 + 0.05(1 - x_i)]y_{i+1} + [-2 + 0.01x_i]y_i + [1 - 0.05(1 - x_i)]y_{i-1} = 0.01x_i.$$

The solution of the corresponding linear system gives

x	0.0	0.1	0.2	0.3	0.4	0.5	0.6
y	0.0000	0.2660	0.5097	0.7357	0.9471	1.1465	1.3353

0.7	0.8	0.9	1.0
1.5149	1.6855	1.8474	2.0000

12. We identify $P(r) = 2/r$, $Q(r) = 0$, $f(r) = 0$, and $h = (4-1)/6 = 0.5$. Then the finite difference equation is

$$\left(1 + \frac{0.5}{r_i}\right) u_{i+1} - 2u_i + \left(1 - \frac{0.5}{r_i}\right) u_{i-1} = 0.$$

The solution of the corresponding linear system gives

r	1.0	1.5	2.0	2.5	3.0	3.5	4.0
u	50.0000	72.2222	83.3333	90.0000	94.4444	97.6190	100.0000

9.R Chapter 9 in Review

3.

x_n	Euler h=0.1	Euler h=0.05	Imp. Euler h=0.1	Imp. Euler h=0.05	RK4 h=0.1	RK4 h=0.05
0.50	0.5000	0.5000	0.5000	0.5000	0.5000	0.5000
0.55		0.5500		0.5512		0.5512
0.60	0.6000	0.6024	0.6048	0.6049	0.6049	0.6049
0.65		0.6573		0.6609		0.6610
0.70	0.7095	0.7144	0.7191	0.7193	0.7194	0.7194
0.75		0.7739		0.7800		0.7801
0.80	0.8283	0.8356	0.8427	0.8430	0.8431	0.8431
0.85		0.8996		0.9082		0.9083
0.90	0.9559	0.9657	0.9752	0.9755	0.9757	0.9757
0.95		1.0340		1.0451		1.0452
1.00	1.0921	1.1044	1.1163	1.1168	1.1169	1.1169

6. The first predictor is $y_3^* = 1.14822731$.

x_n	y_n	
0.0	2.00000000	init. cond.
0.1	1.65620000	RK4
0.2	1.41097281	RK4
0.3	1.24645047	RK4
0.4	1.14796764	ABM

3. If $t = x^3$, then $dt = 3x^2\, dx$ and $x^4\, dx = \frac{1}{3}t^{2/3}\, dt$. Now

$$\int_0^\infty x^4 e^{-x^3}\, dx = \int_0^\infty \frac{1}{3}t^{2/3}e^{-t}\, dt = \frac{1}{3}\int_0^\infty t^{2/3}e^{-t}\, dt$$

$$= \frac{1}{3}\Gamma\left(\frac{5}{3}\right) = \frac{1}{3}(0.89) \approx 0.297.$$

6. For $x > 0$ we integrate by parts:

$$\Gamma(x+1) = \int_0^\infty t^x e^{-t}\, dt$$

$u = t^x$	$dv = e^{-t}\, dt$
$du = xt^{x-1}\, dt$	$v = -e^{-t}$

$$= -t^x e^{-t}\,\bigg|_0^\infty - \int_0^\infty xt^{x-1}(-e^{-t})\, dt$$

$$= x\int_0^\infty t^{x-1}e^{-t}\, dt = x\Gamma(x).$$

A II | Matrices

3. (a) $\mathbf{AB} = \begin{pmatrix} -2-9 & 12-6 \\ 5+12 & -30+8 \end{pmatrix} = \begin{pmatrix} -11 & 6 \\ 17 & -22 \end{pmatrix}$

(b) $\mathbf{BA} = \begin{pmatrix} -2-30 & 3+24 \\ 6-10 & -9+8 \end{pmatrix} = \begin{pmatrix} -32 & 27 \\ -4 & -1 \end{pmatrix}$

(c) $\mathbf{A}^2 = \begin{pmatrix} 4+15 & -6-12 \\ -10-20 & 15+16 \end{pmatrix} = \begin{pmatrix} 19 & -18 \\ -30 & 31 \end{pmatrix}$

(d) $\mathbf{B}^2 = \begin{pmatrix} 1+18 & -6+12 \\ -3+6 & 18+4 \end{pmatrix} = \begin{pmatrix} 19 & 6 \\ 3 & 22 \end{pmatrix}$

6. (a) $\mathbf{AB} = (5 \quad -6 \quad 7) \begin{pmatrix} 3 \\ 4 \\ -1 \end{pmatrix} = (-16)$

(b) $\mathbf{BA} = \begin{pmatrix} 3 \\ 4 \\ -1 \end{pmatrix} (5 \quad -6 \quad 7) = \begin{pmatrix} 15 & -18 & 21 \\ 20 & -24 & 28 \\ -5 & 6 & -7 \end{pmatrix}$

(c) $(\mathbf{BA})\mathbf{C} = \begin{pmatrix} 15 & -18 & 21 \\ 20 & -24 & 28 \\ -5 & 6 & -7 \end{pmatrix} \begin{pmatrix} 1 & 2 & 4 \\ 0 & 1 & -1 \\ 3 & 2 & 1 \end{pmatrix} = \begin{pmatrix} 78 & 54 & 99 \\ 104 & 72 & 132 \\ -26 & -18 & -33 \end{pmatrix}$

(d) Since $\mathbf{AB}$ is 1×1 and $\mathbf{C}$ is 3×3 the product $(\mathbf{AB})\mathbf{C}$ is not defined.

9. (a) $(\mathbf{AB})^T = \begin{pmatrix} 7 & 10 \\ 38 & 75 \end{pmatrix}^T = \begin{pmatrix} 7 & 38 \\ 10 & 75 \end{pmatrix}$

(b) $\mathbf{B}^T\mathbf{A}^T = \begin{pmatrix} 5 & -2 \\ 10 & -5 \end{pmatrix}\begin{pmatrix} 3 & 8 \\ 4 & 1 \end{pmatrix} = \begin{pmatrix} 7 & 38 \\ 10 & 75 \end{pmatrix}$

12. $\begin{pmatrix} 6t \\ 3t^2 \\ -3t \end{pmatrix} + \begin{pmatrix} -t+1 \\ -t^2+t \\ 3t-3 \end{pmatrix} - \begin{pmatrix} 6t \\ 8 \\ -10t \end{pmatrix} = \begin{pmatrix} -t+1 \\ 2t^2+t-8 \\ 10t-3 \end{pmatrix}$

15. Since $\det \mathbf{A} = 0$, $\mathbf{A}$ is singular.

18. Since $\det \mathbf{A} = -6$, $\mathbf{A}$ is nonsingular.

$$\mathbf{A}^{-1} = -\frac{1}{6}\begin{pmatrix} 2 & -10 \\ -2 & 7 \end{pmatrix}$$

21. Since $\det \mathbf{A} = -9$, $\mathbf{A}$ is nonsingular. The cofactors are

$$\begin{array}{lll} A_{11} = -2 & A_{12} = -13 & A_{13} = 8 \\ A_{21} = -2 & A_{22} = 5 & A_{23} = -1 \\ A_{31} = -1 & A_{32} = 7 & A_{33} = -5. \end{array}$$

Then

$$\mathbf{A}^{-1} = -\frac{1}{9}\begin{pmatrix} -2 & -13 & 8 \\ -2 & 5 & -1 \\ -1 & 7 & -5 \end{pmatrix}^T = -\frac{1}{9}\begin{pmatrix} -2 & -2 & -1 \\ -13 & 5 & 7 \\ 8 & -1 & -5 \end{pmatrix}.$$

24. Since $\det \mathbf{A}(t) = 2e^{2t} \neq 0$, $\mathbf{A}$ is nonsingular.

$$\mathbf{A}^{-1} = \frac{1}{2}e^{-2t}\begin{pmatrix} e^t\sin t & 2e^t\cos t \\ -e^t\cos t & 2e^t\sin t \end{pmatrix}$$

27. $\mathbf{X} = \begin{pmatrix} 2e^{2t} + 8e^{-3t} \\ -2e^{2t} + 4e^{-3t} \end{pmatrix}$ so that $\dfrac{d\mathbf{X}}{dt} = \begin{pmatrix} 4e^{2t} - 24e^{-3t} \\ -4e^{2t} - 12e^{-3t} \end{pmatrix}.$

30. (a) $\dfrac{d\mathbf{A}}{dt} = \begin{pmatrix} -2t/(t^2+1)^2 & 3 \\ 2t & 1 \end{pmatrix}$

(b) $\dfrac{d\mathbf{B}}{dt} = \begin{pmatrix} 6 & 0 \\ -1/t^2 & 4 \end{pmatrix}$

(c) $\displaystyle\int_0^1 \mathbf{A}(t)\,dt = \begin{pmatrix} \tan^{-1} t & \frac{3}{2}t^2 \\ \frac{1}{3}t^3 & \frac{1}{2}t^2 \end{pmatrix}\Bigg|_{t=0}^{t=1} = \begin{pmatrix} \frac{\pi}{4} & \frac{3}{2} \\ \frac{1}{3} & \frac{1}{2} \end{pmatrix}$

(d) $\displaystyle\int_1^2 \mathbf{B}(t)\,dt = \begin{pmatrix} 3t^2 & 2t \\ \ln t & 2t^2 \end{pmatrix}\Bigg|_{t=1}^{t=2} = \begin{pmatrix} 9 & 2 \\ \ln 2 & 6 \end{pmatrix}$

(e) $\mathbf{A}(t)\mathbf{B}(t) = \begin{pmatrix} 6t/(t^2+1)+3 & 2/(t^2+1)+12t^2 \\ 6t^3+1 & 2t^2+4t^2 \end{pmatrix}$

(f) $\dfrac{d}{dt}\mathbf{A}(t)\mathbf{B}(t) = \begin{pmatrix} (6-6t^2)/(t^2+1)^2 & -4t/(t^2+1)^2+24t \\ 18t^2 & 12t \end{pmatrix}$

(g) $\displaystyle\int_1^t \mathbf{A}(s)\mathbf{B}(s)\,ds = \int_1^t \begin{pmatrix} 6s/(s^2+1)+3 & 2/(s^2+1)+12s^2 \\ 6s^3+1 & 6s^2 \end{pmatrix}\,ds$

$$= \begin{pmatrix} 3\ln(t^2+1)+3t-3\ln 2-3 & 2\tan^{-1}t+4t^3-\pi/2-4 \\ 3t^4/2+t-5/2 & 2t^3-2 \end{pmatrix}$$

33. $\begin{pmatrix} 1 & -1 & -5 & | & 7 \\ 5 & 4 & -16 & | & -10 \\ 0 & 1 & 1 & | & -5 \end{pmatrix} \Longrightarrow \begin{pmatrix} 1 & -1 & -5 & | & 7 \\ 0 & 1 & 1 & | & -5 \\ 0 & 9 & 9 & | & -45 \end{pmatrix} \Longrightarrow \begin{pmatrix} 1 & 0 & -4 & | & 2 \\ 0 & 1 & 1 & | & -5 \\ 0 & 0 & 0 & | & 0 \end{pmatrix}$

Letting $z = t$ we find $y = -5 - t$, and $x = 2 + 4t$.

36.
$$\begin{pmatrix} 1 & 0 & 2 & | & 8 \\ 1 & 2 & -2 & | & 4 \\ 2 & 5 & -6 & | & 6 \end{pmatrix} \implies \begin{pmatrix} 1 & 0 & 2 & | & 8 \\ 0 & 2 & -4 & | & -4 \\ 0 & 5 & -10 & | & -10 \end{pmatrix} \implies \begin{pmatrix} 1 & 0 & 2 & | & 8 \\ 0 & 1 & -2 & | & -2 \\ 0 & 0 & 0 & | & 0 \end{pmatrix}$$

Letting $z = t$ we find $y = -2 + 2t$, and $x = 8 - 2t$.

39.
$$\begin{pmatrix} 1 & 2 & 4 & | & 2 \\ 2 & 4 & 3 & | & 1 \\ 1 & 2 & -1 & | & 7 \end{pmatrix} \implies \begin{pmatrix} 1 & 2 & 4 & | & 2 \\ 0 & 0 & -5 & | & -3 \\ 0 & 0 & -5 & | & 5 \end{pmatrix} \implies \begin{pmatrix} 1 & 2 & 0 & | & -2/5 \\ 0 & 0 & 1 & | & 3/5 \\ 0 & 0 & 0 & | & 8 \end{pmatrix}$$

There is no solution.

42.
$$\begin{bmatrix} 2 & 4 & -2 & | & 1 & 0 & 0 \\ 4 & 2 & -2 & | & 0 & 1 & 0 \\ 8 & 10 & -6 & | & 0 & 0 & 1 \end{bmatrix} \xrightarrow[\text{operations}]{\text{row}} \begin{bmatrix} 1 & 2 & -1 & | & \frac{1}{2} & 0 & 0 \\ 0 & 1 & -\frac{1}{3} & | & \frac{1}{3} & -\frac{1}{6} & 0 \\ 0 & 0 & 0 & | & -2 & -1 & 1 \end{bmatrix} ; \quad \mathbf{A} \text{ is singular.}$$

45.
$$\begin{bmatrix} 1 & 2 & 3 & 1 & | & 1 & 0 & 0 & 0 \\ -1 & 0 & 2 & 1 & | & 0 & 1 & 0 & 0 \\ 2 & 1 & -3 & 0 & | & 0 & 0 & 1 & 0 \\ 1 & 1 & 2 & 1 & | & 0 & 0 & 0 & 1 \end{bmatrix} \xrightarrow[\text{operations}]{\text{row}} \begin{bmatrix} 1 & 2 & 3 & 1 & | & 1 & 0 & 0 & 0 \\ 0 & 1 & \frac{5}{2} & 1 & | & \frac{1}{2} & \frac{1}{2} & 0 & 0 \\ 0 & 0 & 1 & -\frac{2}{3} & | & \frac{1}{3} & -1 & -\frac{2}{3} & 0 \\ 0 & 0 & 0 & 1 & | & -\frac{1}{2} & 1 & \frac{1}{2} & \frac{1}{2} \end{bmatrix}$$

$$\xrightarrow[\text{operations}]{\text{row}} \begin{bmatrix} 1 & 0 & 0 & 0 & | & -\frac{1}{2} & -\frac{2}{3} & -\frac{1}{6} & \frac{7}{6} \\ 0 & 1 & 0 & 0 & | & 1 & \frac{1}{3} & \frac{1}{3} & -\frac{4}{3} \\ 0 & 0 & 1 & 0 & | & 0 & -\frac{1}{3} & -\frac{1}{3} & \frac{1}{3} \\ 0 & 0 & 0 & 1 & | & -\frac{1}{2} & 1 & \frac{1}{2} & \frac{1}{2} \end{bmatrix} ; \quad \mathbf{A}^{-1} = \begin{bmatrix} -\frac{1}{2} & -\frac{2}{3} & -\frac{1}{6} & \frac{7}{6} \\ 1 & \frac{1}{3} & \frac{1}{3} & -\frac{4}{3} \\ 0 & -\frac{1}{3} & -\frac{1}{3} & \frac{1}{3} \\ -\frac{1}{2} & 1 & \frac{1}{2} & \frac{1}{2} \end{bmatrix}$$

48. We solve

$$\det(\mathbf{A} - \lambda\mathbf{I}) = \begin{vmatrix} 2 - \lambda & 1 \\ 2 & 1 - \lambda \end{vmatrix} = \lambda(\lambda - 3) = 0.$$

For $\lambda_1 = 0$ we have

$$\begin{pmatrix} 2 & 1 & | & 0 \\ 2 & 1 & | & 0 \end{pmatrix} \implies \begin{pmatrix} 1 & 1/2 & | & 0 \\ 0 & 0 & | & 0 \end{pmatrix}$$

so that $k_1 = -\frac{1}{2}k_2$. If $k_2 = 2$ then

$$\mathbf{K}_1 = \begin{pmatrix} -1 \\ 2 \end{pmatrix}.$$

For $\lambda_2 = 3$ we have

$$\begin{pmatrix} -1 & 1 & | & 0 \\ 2 & -2 & | & 0 \end{pmatrix} \implies \begin{pmatrix} 1 & -1 & | & 0 \\ 0 & 0 & | & 0 \end{pmatrix}$$

so that $k_1 = k_2$. If $k_2 = 1$ then

$$\mathbf{K}_2 = \begin{pmatrix} 1 \\ 1 \end{pmatrix}.$$

51. We solve

$$\det(\mathbf{A} - \lambda\mathbf{I}) = \begin{vmatrix} 5 - \lambda & -1 & 0 \\ 0 & -5 - \lambda & 9 \\ 5 & -1 & -\lambda \end{vmatrix} = \lambda(4 - \lambda)(\lambda + 4) = 0.$$

If $\lambda_1 = 0$ then

$$\begin{pmatrix} 5 & -1 & 0 & | & 0 \\ 0 & -5 & 9 & | & 0 \\ 5 & -1 & 0 & | & 0 \end{pmatrix} \implies \begin{pmatrix} 1 & 0 & -9/25 & | & 0 \\ 0 & 1 & -9/5 & | & 0 \\ 0 & 0 & 0 & | & 0 \end{pmatrix}$$

so that $k_1 = \frac{9}{25}k_3$ and $k_2 = \frac{9}{5}k_3$. If $k_3 = 25$ then

$$\mathbf{K}_1 = \begin{pmatrix} 9 \\ 45 \\ 25 \end{pmatrix}.$$

If $\lambda_2 = 4$ then

$$\begin{pmatrix} 1 & -1 & 0 & | & 0 \\ 0 & -9 & 9 & | & 0 \\ 5 & -1 & -4 & | & 0 \end{pmatrix} \implies \begin{pmatrix} 1 & 0 & -1 & | & 0 \\ 0 & 1 & -1 & | & 0 \\ 0 & 0 & 0 & | & 0 \end{pmatrix}$$

so that $k_1 = k_3$ and $k_2 = k_3$. If $k_3 = 1$ then

$$\mathbf{K}_2 = \begin{pmatrix} 1 \\ 1 \\ 1 \end{pmatrix}.$$

If $\lambda_3 = -4$ then

$$\begin{pmatrix} 9 & -1 & 0 & | & 0 \\ 0 & -1 & 9 & | & 0 \\ 5 & -1 & 4 & | & 0 \end{pmatrix} \implies \begin{pmatrix} 1 & 0 & -1 & | & 0 \\ 0 & 1 & -9 & | & 0 \\ 0 & 0 & 0 & | & 0 \end{pmatrix}$$

so that $k_1 = k_3$ and $k_2 = 9k_3$. If $k_3 = 1$ then

$$\mathbf{K}_3 = \begin{pmatrix} 1 \\ 9 \\ 1 \end{pmatrix}.$$

54. We solve

$$\det(\mathbf{A} - \lambda\mathbf{I}) = \begin{vmatrix} 1-\lambda & 6 & 0 \\ 0 & 2-\lambda & 1 \\ 0 & 1 & 2-\lambda \end{vmatrix} = (3-\lambda)(1-\lambda)^2 = 0.$$

For $\lambda = 3$ we have

$$\begin{pmatrix} -2 & 6 & 0 & | & 0 \\ 0 & 0 & 0 & | & 0 \\ 0 & 1 & -1 & | & 0 \end{pmatrix} \implies \begin{pmatrix} 1 & 0 & -3 & | & 0 \\ 0 & 1 & -1 & | & 0 \\ 0 & 0 & 0 & | & 0 \end{pmatrix}$$

so that $k_1 = 3k_3$ and $k_2 = k_3$. If $k_3 = 1$ then

$$\mathbf{K}_1 = \begin{pmatrix} 3 \\ 1 \\ 1 \end{pmatrix}.$$

For $\lambda_2 = \lambda_3 = 1$ we have

$$\begin{pmatrix} 0 & 6 & 0 & | & 0 \\ 0 & 1 & 1 & | & 0 \\ 0 & 1 & 1 & | & 0 \end{pmatrix} \implies \begin{pmatrix} 0 & 1 & 0 & | & 0 \\ 0 & 0 & 1 & | & 0 \\ 0 & 0 & 0 & | & 0 \end{pmatrix}$$

so that $k_2 = 0$ and $k_3 = 0$. If $k_1 = 1$ then

$$\mathbf{K}_2 = \begin{pmatrix} 1 \\ 0 \\ 0 \end{pmatrix}.$$

57. Let

$$\mathbf{A} = \begin{pmatrix} a_{11} & a_{12} \\ a_{21} & a_{22} \end{pmatrix}.$$

Then

$$\frac{d}{dt}[\mathbf{A}(t)\mathbf{X}(t)] = \frac{d}{dt}\begin{pmatrix} a_1 & a_2 \\ a_3 & a_4 \end{pmatrix}\begin{pmatrix} x_1 \\ x_2 \end{pmatrix} = \frac{d}{dt}\begin{pmatrix} a_1x_1 + a_2x_2 \\ a_3x_1 + a_4x_2 \end{pmatrix} = \begin{pmatrix} a_1x_1' + a_1'x_1 + a_2x_2' + a_2'x_2 \\ a_3x_1' + a_3'x_1 + a_4x_2' + a_4'x_2 \end{pmatrix}$$

$$= \begin{pmatrix} a_1 & a_2 \\ a_3 & a_4 \end{pmatrix}\begin{pmatrix} x_1' \\ x_2' \end{pmatrix} + \begin{pmatrix} a_1' & a_2' \\ a_3' & a_4' \end{pmatrix}\begin{pmatrix} x_1 \\ x_2 \end{pmatrix} = \mathbf{A}(t)\mathbf{X}'(t) + \mathbf{A}'(t)\mathbf{X}(t).$$

60. Since

$$(\mathbf{AB})(\mathbf{B}^{-1}\mathbf{A}^{-1}) = \mathbf{A}(\mathbf{BB}^{-1})\mathbf{A}^{-1} = \mathbf{AIA}^{-1} = \mathbf{AA}^{-1} = \mathbf{I}$$

and

$$(\mathbf{B}^{-1}\mathbf{A}^{-1})(\mathbf{AB}) = \mathbf{B}^{-1}(\mathbf{A}^{-1}\mathbf{A})\mathbf{B} = \mathbf{B}^{-1}\mathbf{IB} = \mathbf{B}^{-1}\mathbf{B} = \mathbf{I}$$

we have

$$(\mathbf{AB})^{-1} = \mathbf{B}^{-1}\mathbf{A}^{-1}.$$

CPSIA information can be obtained
at www.ICGtesting.com
Printed in the USA
FFOW020655310113
796FF

9 781133 491927